on What is
a freedom enchantment in 23 ACTs

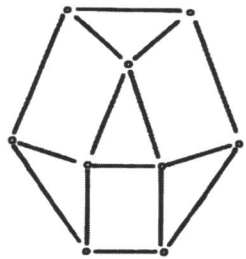

from **ja wallin**

Some Other Titles From New Falcon Publications

Aha! The Sevenfold Mystery of the Ineffable Love — **By Aleister Crowley**
Bio-Etheric Healing — **By Trudy Lanitis**
Undoing Yourself With Energized Meditation and Other Devices
Secrets of Western Tantra: The Sexuality of the Middle Path
Dogma Daze — **By Christopher S. Hyatt, Ph.D.**
Rebels & Devils; The Psychology of Liberation **Edited by Christopher S. Hyatt, Ph.D.**
Aleister Crowley's Illustrated Goetia
Taboo: Sex, Religion & Magick
Sex Magic, Tantra & Tarot: The Way of the Secret Lover
  By Christopher S. Hyatt, Ph.D., and Lon Milo DuQuette
Pacts With The Devil
Urban Voodoo: A Beginner's Guide to Afro-Caribbean Magic
  By Jason Black and Christopher S. Hyatt, Ph.D.
The Psychopath's Bible — **By Christopher S. Hyatt, Ph.D., and Jack Willis**
Ask Baba Lon — **By Lon Milo DuQuette**
Aleister Crowley and the Treasure House of Images **By J.F.C. Fuller, Aleister Crowley, Lon Milo DuQuette and Nancy Wasserman**
Enochian World of Aleister Crowley **By Lon Milo DuQuette and Aleister Crowley**

Info-Psychology   Neuropolitique   The Game of Life
What Does WoMan Want? — **By Timothy Leary, Ph.D.**

Be Yourself - A Guide to Relaxation and Health
Dr. Israel Regardie's Definitive Work on Aleister Crowley, The Eye In The Triangle
Healing Energy, Prayer and Relaxation
My Rosicrucian Adventure
Teachers of Fulfillment
The Complete Golden Dawn System of Magic
The Eye in the Triangle: An Interpretation of Aleister Crowley
The Golden Dawn Audio CDs
The Legend of Aleister Crowley
The Portable Complete Golden Dawn System of Magic
The Tree of Life
What You Should Know About the Golden Dawn — **By Dr. Israel Regardie**

Roll Away The Stone/The Herb Dangerous **By Israel Regardie and Aleister Crowley**

Rebellion, Revolution and Religiousness — **By Osho**
Reichian Therapy: A Practical Guide for Home Use — **By Dr. Jack Willis**
Woman's Orgasm: A Guide to Sexual Satisfaction **By Benjamin Graber, M.D., and Georgia Kline-Graber, R.N.**
Shaping Formless Fire   Seizing Power   Taking Power — **By Stephen Mace**
The Illuminati Conspiracy: The Sapiens System — **By Donald Holmes, M.D.**
The Secret Inner Order Rituals of the Golden Dawn — **By Pat Zalewski**

MANY OF OUR TITLES AVAILABLE ON KINDLE!
Please visit our website at http://www.newfalcon.com

on What is
a freedom enchantment in 23 ACTs

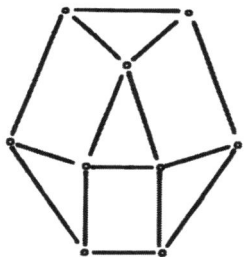

from **ja wallin**

NEW FALCON PUBLICATIONS
LAS VEGAS, NEVADA, U.S.A.

Copyright © 2018 J. F. Nystrom

All rights reserved. No part of this book,
in part or in whole, may be reproduced, transmitted,
or utilized, in any form or by any means, electronic or mechanical,
including photocopying, recording, or by any information storage
and retrieval system, without permission in writing
from the publisher, except for brief quotations
in critical articles, books and reviews.

ISBN 13: 978-1-56184-542-2
ISBN 10: 1-56184-542-6

First Edition 2018

The paper used in this publication meets the minimum requirements
of the American National Standard for Permanence of
Paper for Printed Library Materials Z39.48-1984

Printed in USA

NEW FALCON PUBLICATIONS
9550 South Eastern Avenue • Suite 253
Las Vegas, NV 89123
www.newfalcon.com
email: info@newfalcon.com

dedication
Of the 23 ACTs on What is:

to the theurgists, and others who do not submit to the sons of Belial (those sOb's); let the divine shine today, let tomorrow bring what it may.

For the theurgists, and others who oppose the machinations of the sOb's; let the(se) seeking foul(s continue on their path in safety.

in accordance with the theurgists' will; let the divine spirit provide counsel to all things, along their course of destiny, in association with, necessity.

With respect,

herb the theurgist
hrbththrst
hrbthrst

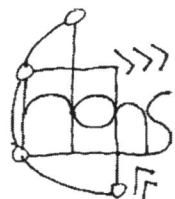

oVerture

as Any two walk a path in a meadow, what is it that they are allowed to discuss? Is this scene independent of time, which is to say, are concerns universal throughout the ages? Be advised that a discussion presumes a language does it not, or can we 'speak' to another with thoughts without words?

more Questions, more questions; but, again, can any question be asked? Is not this the same as to what is allowed to be discussed? When a question is posed between only two, whom else is privy to the query?

can One ask about a 'thing' that they k-now not of? Is it better to ask only about what one already k-nows? Better or worse, but be careful of what is allowed. To ponder the nature of universals one finds in their surroundings would seem an inalienable right, but can it be justified to police such activity and allow only certain thoughts – dictate then in a sense what can occur within the individuals plight?

over Many aeons and situations involving wo/man, all is in fact not allowed. For as long as we k-now, it is only the priests whom are allowed, to ponder freely into the nature of wo/man. so As we move on, always maintaining civility, we say together now, 'damn Zeus's tyranny,' for we will think about what we want, as we all strive to help our fellow wo/man, yes, damn Zeus's tyranny on the affairs of wo/man.

be Warned however, and beware of the wrath, for whatever the age, the seekers of truth are always cut off at the pass. Take a lesson of what happened to one of their gods, ole Prometheus came and taught the nature of things to wo/man, and Zeus put him in chains to be eaten everyday.

so If you aspire to enlighten your fellow wo/man, or be an archetypal 'catcher in the rye,' ye k-now what to expect from the powers that be, the cabal of maniac magicians, they do not 'like' you or me. Here then a freedom enchantment four those who wood dare. Be thus prepared, to welcome fate, as it all indeed may be written, or may not yet have taken place. For we k-now not the nature of time nor of space, and damn Zeus's tyranny, personally eye welcome (the goddess) Fate[1].

---

[1] *anonymous Domesticated primate 1* (aDp1): what Would Robert Antom Wilson want to add to this oVerture, if you could divine such a thing?
aDp2: he May (or June) thunk it all quite absurd!
aDp1: it Is of course absurd to believe any wo/man could ever truly k-now, the True nature of things we see all around.
aDp2: but Should we dare challenge the absurd, and continue to endeavor to learn the best we can; the nature of things, and the nature of wo/man?
aDp1: cadmus Would call us absurdist, if he happened to give us a thought, absurdist all of us who would dare continue on after realizing we could never accomplish our tasks.
aDp2: in Certain times, it is told, it was called a magician, who will be so bold, to truly investigate the nature of the physical and dream worlds, for it all could be so scary, to not do what one is told.
aDp1: ah, The magician is truly the name for us all, but do not forget then what can be done, for The magician who would mess with the 'will' of their fellow wo/man is for sure a sorcerer (or sorceress), and thus is committed until they would pronounce their regret. Divine magic then, properly called theurgy, is the magic one would use to help one's immortal soul, and to attempt to make sense of the absurd world in which they reside.
aDp2: can All partake of the divine magic that is available to all, or is the deception so powerful that only a few can choose, to be an absurdist in this funny little place some call our world?
aDp1: ah, Then a question we would ask of a / the god(s), shall I do that tonight in my dreams, or should I abandon this task; of seeking to k-now about the nature of wo/man?
aDp2: robert Anton Wilson would say, since you previously asked, that if you want to work for the Greater Poop, listen to what no one has to say. Ney, he would enlighten us also to search for the goddess, or check out that little book on how he found the goddess
aDp1: ah Yes, and what he did to her when he found her; it is all quite absurd, but lets pretend we have his and others blessings, to continue our tasks; and thus damn Zeus's tyranny, and let the seeking fools continue on their path
aDp2: so, Finally at last, lets point our adventurers to pages 17 and 18 of that little book
aDp1: the One about the goddess Eris?
aDp2: yes, You silly absurdist, you aspiring theurgist, of course thats the book
aDp1: ah, The story then of how Zeus did not want Eris at the party, for she was a trouble maker (and would not be allowed to dance)
aDp2: and What did she do?
aDp1: she Fashioned an apple of Discord, for the most beautiful of all
aDp2: and Then a battle ensued, all due in fact to this Original snub
aDp1: and All us domesticated primates must now all pay, because Zeus did not want Eris about
aDp2: lets Stop now, please, and ack-nowledge it is all quite absurd, and agree
to end this by shouting together, at the top of our lungs
domesticated Primates (together in song): DAMN ZEUS'S TYRANNY, DAMN ZEUS'S TYRANNY, DAMN ZEUS'S TYRANNY ON THE AFFAIRS OF WO/MAN

ACT I

epic Of herb the theurgist the Computer journey To the east amerikan Gypsy taking The medicine faster Than any animal can run the Agricultural and mechanical college the World's friendly genius the Geometry of nature old Places old Times 17 Years in the wilderness into The new world order are You living in a computer simulation? what Is a computer? the Heresy of scientific materialism physical Universe needs a non-physical what Descartes supposed let Me get it back baby Where i come from computational Cosmography and all that that entails witch Will it be onward Towards a need for theurgy hmmm

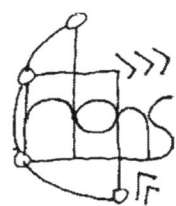

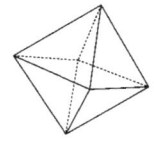

INTER-MISSION

once Upon a path theurgist Of atlantis

epic Of herb the theurgist

> Even at first reading, I could see that Korzybski had the answer to at least one question that had perplexed me for years – namely, What is "reality?" According to Korzybski, the only correct way to answer the question begins with recognizing that "reality" is – a word.
> - RAW

herb was born in a mountainousness area in the northwest of a country that at this time was seeking world around hegemony. This was of course just a continuation of the age old struggle between the Divine and the darkness - a struggle that not only appears in the overall affairs of man (most likely since one of the great cataclysms), but within each man[2] also. This particular country at this time was involved in the now well-known ruse of pretending to be the bringer of Light to the planet, while in action playing out deeds that made it a global center of darkness. herb was born 533 moons after his father, in the same town, right across a street from where his father was born into this world. This town has a rough river rolling through it, that contains fish that can live longer than a man.

It was a time on the planet when the whole world was connected via a mysterious fluid type magical substance called electricity - which animated, in its way, all the machinery needed for these aforementioned global connections. The use of "electricity" combined with ingenious machine making techniques had allowed *humans* (and you k-now who you are) to become self-proclaimed *Masters of the planet*.

As herb was just recently born into this world, he consciously was not aware of this overall situation, and in the earliest of his

---

[2] "man" is used as gender neutral designation for humans - and you k-now who you are - throughout the epic. wo/man should be read as 'woman and man.'

years depended solely on his loving parents and (as what he would come to k-now as) universal providence. It was in the second Earth year of his life that the family moved to a place on the west coast of this hegemonic country, and it was there that they set up house in a valley that would become famous for being the center of the continual development of what was then called a *computer*.

the Computer

A computer is a machine that can *remember* things (via a memory system) and that can carry out instructions. There were currently on the planet at this time many different types of computers. The so-called analog computers would work with *continuous* values of quantities and then effect actions based on the instructions that were *programmed* into the machine via certain types of circuitry. The term *circuitry* refers to organizations of elements that are meant to control the flow of that mysterious fluid type magical substance called electricity.

The most famous type of computer at this time however was the so-called digital computer. The digital computer (which from now on I will call simply the computer) was at this time based on the manipulation of two distinct abstract values. The two distinct values that the computer used were designated as 1 (one) and 0 (zero). When the distinct values are represented this way, binary mathematics, which is a base 2 mathematics, can be used to describe the internal workings of the computer. Other common designations for these two distinct values at the center of computational decision-making was ON/OFF, and HIGH/LOW. The terminology HIGH/LOW is very descriptive of how the circuitry of the digital computer worked at this time, in that the digital circuitry worked with two distinct levels of *voltage* (which is one of the measurement values typically used when designing circuitry to control that mysterious fluid type magical substance called electricity).

The usefulness of computers in the earliest days was for calculating the results of mathematical equations. Thus, the very earliest computers replaced calculations that were being done by hand, or via mechanical calculating devices. Some of these early computers were used to calculate artillery tables for warships that rode on the oceans and seas of the planet, and shot projectiles from the warship to both sea-bound and land-based targets. The equations that produced the many different artillery tables were more involved than the projectile motion equations typically taught early in a scientist's education at the time, due to the fact that planet is spinning about an axis, and thus any projectile will actually deviate from what one might call a straight line trajectory. To wit, in the Northern hemisphere of the planet, the trajectory of the projectile deflects slightly to the right, while in the Southern hemisphere of the planet the projectile deflects slightly to the left; all due to the so-called Coriolis force (which accounts for the aforementioned fact that the mathematics needs to accommodate the fact that the surface of the planet is spinning on an axis with a certain angular velocity). The use of automated calculation for these artillery tables improved both accuracy of the calculations and speed of production of said tables.

Other uses of the digital computer soon became apparent when trying to solve, for example, so-called combinatorial problems (for example, how best to pick one thing from the very many options), many numerical problems, or really any problem for which a solution could be described by an *algorithm*. An algorithm is a type of recipe (if you will), and just as with a cooking recipe, an algorithm will list all the preconditions (the recipes ingredients), and then all the needed steps to come to a solution (the recipes finished food item). Algorithms can be written for virtually any process that has a fixed number of requisite conditions and a finite number of operations to obtain the result. Example algorithms include one for the ancient game of tic-tac-toe guaranteed to give the user (of said algorithm) a draw or a win depending only on the initial placement of the opposing player; and the ancient algorithm for obtaining the square root of a number attributed to the great shaman Newton.

The management of monies and statistics was then soon also addressed by computing machines. Although there had been tabulating machines around for ages before the *electronic* digital computer (so-called electronic because the operation of said machines depends on manipulation of that mysterious fluid type magical substance called electricity), the advent of the computer soon replaced the tabulating by hand of all the monies and statistics work with algorithms and memory organization techniques for these new computational machines. Also at this time, the Powers That Be[3] began implementation of computational techniques designed to track every aspect of almost every persons lives. Soon it would be common place for cameras and microphones to be installed all about most cities, and these surveillance devices would be hooked into the computers to give them the power to monitor, discredit and/or analyze the behavior of any common wo/man.

journey To the east

Soon after herb's sister was born into the world, the family moved from the west coast to the east coast of the hegemonic country into which herb was born. herb's father at this time was working for the largest makers of computers on the planet, and as such, this company was very instrumental in providing computational machines that supported the hegemonic plans of the country that herb was born into.

During herb's early schooling it became readily apparent that there was not a mathematics that he could not learn and understand. Even though at this time in this hegemonic country of his birth, the planned destruction of education was well under way,

---

[3] throughout the epic one can substitute either (*) the cabal of maniac magicians, or (*) the sons of Belial (those sOb's), for the Powers That Be.

herb was able to get a quality overview of many of the basic principles of the then modern mathematics. herb also got good introductions to the sciences of those days, including the physics, biology, and chemistry fields. herb did not k-now at that time that there was also a lot of hidden science being developed and utilized in the deep dark secret laboratories of the cabal of maniac magicians – funded, monitored, and staffed in part by the sons of Belial (those sOb's).

During these early years, herb was introduced to the organized religion of the time, and taken to church almost every Sun-day by his mother. The specific church herb attended was affiliated with the largest organized religion in the world at that time. Even in his young years, herb thought some things odd about the whole affair. When sermons at the church talked about being civil to others, he had no disagreement with that, but what about the rituals during mass that simulated cannibalism and vampirism wherein the believer is to image eating the body and drinking the blood of the Christ - odd indeed, no matter what anyone says about it. Before being *confirmed* into the religion (one of the many, as it seems, associated with / organized by the sOb's) wherein the young person (supposedly) had a choice, herb and a friend did determine that they would choose to pass on the opportunity to be confirmed if not for the fact that their mothers would be heart-broken; thus both herb and his friend went through with the ceremony that, I suppose, initiated the both of them into this now all too famous ancient blood-cult. It became well-known in due course that this particular church was also part of that ruse of pretending to be the bringer of the Divine to the planet, while in action playing out deeds (mostly in a secret and occulted manner) that made it a global center of darkness.[4]

---

[4] Later in life herb would come to understand the role played by the global Banksters also, and then come to understand more completely the danger posed to humanities plight by the UnHoly trinity of government / organized religion / global Banksters, and how this Unholy trinity had been manipulating events (and history) for so so many years.

During these years in the east, herb was at an age where he could judge for himself the merits of peoples of different races, as this was a location where representatives of many different tribes lived. herb had friends and acquaintances from all the different groups. At the same time however herb could sense that all was not right in this hegemonic country of his birth, where it seemed the sOb's were always trying to fabricate one reason after another for peoples from different tribes to hate, rather than love, each other.

amerikan Gypsy

In his early teen years, herb's family moved again, this time to the famous region of the planet known as Tejas (then still part of the hegemonic country of his birth). As this was already the fourth different place herb had lived at still such a young age, he would often tell the tale later of how he was an *amerikan Gypsy*. It was only later in life that he would understand how true to providence this statement was, as it was predominately the gypsies of the planet at that time who still maintained historical memories closest to the real facts. It was the case in these days, as in almost all ancient times, that history as taught to most children (and adults) was for the most part a fabrication and/or bastardization of real events - which is one of the well-known techniques employed in the battle between the Divine and dark forces operating throughout all ages. This suppression and corruption of history has always been a key element for control of general populations.

herb continued to excel at mathematics and the physical sciences, and he also became interested in philosophy. As herb's body was changing into a young adult, the amorous issues that arise for these young people also was coming into play. Impacting herb's understanding of all this was the fact that the society of the hegemonic country into which herb was born was being turned upside down for young people at the time; as in this time everyone had access to *television* (or tv). The television was a communication device that displayed images and produced sounds into the houses and gathering places of humans on the planet. This television device was powered by that mysterious fluid type magical substance called electricity, and the content of the television *message* at this time was sent through the air of the planet (or through some other transmission media - typically involving wires, or, in later years, via the so-called fiber optics) using *electromagnetic waves* and elaborate encoding techniques so that the images and sounds could be transmitted via these waves. It was k-nown at that time that the electrical power that ran many devices, and the electromagnetic waves used to transmit signals, were of the same nature, i.e., part of that mysterious fluid type magical substance called electricity. From the very first televisions, the sons of Belial (those sOb's) used these devices to spread a propaganda message that tried to convince all viewers (the name for someone watching a television) about the nature of their existence, and also how they should behave given this false narrative on the nature of existence. The television was thus a powerful control tool in the age old struggle between the Divine and the darkness.

It was also during this time in his schooling that herb began to write *computer programs*. A computer program is a way to write out a series of instructions for a computer to follow. The *programmer* will use an algorithm for a problem and then write out a series of instructions in a specific *computer language*. There were many computer languages available at this time; and once a program was written in a particular language, another program would be invoked that converted the program into another (low-level) program that the computer would actually run. Just as with mathematics, herb was able to understand almost everything about programming as his education continued. herb was also able to intuit how computer output could be used for deception. It was during a class in how governments (supposedly!) operate that herb wrote a program that counterfeited the class printouts for various countries resources in order for his particular country to look more prosperous and powerful than it actually was. This ruse worked for a short time before some of the other players in this game of government started paying a bit of attention and eventually saw through the false statements that were being produced by herb's (intentionally) deceptive programs.

Later in life herb would learn of predictive programming techniques brought about through psychological research and sophisticated propaganda efforts that would have a tremendous effect on the average humans ability to think beyond simple stimulus-response ideas. This type of massive deception with the aid of the computer became very widespread, and as herb became to see the reality of its usefulness to the sOb's, herb continued to hope (but not yet enchant) that all humans would soon be able to see through this ruse being run by what herb began to accurately call the *cabal of maniac magicians* (i.e., those sOb's).

taking The medicine

It was during these teen years that herb first took a popular medicine of the time for people who (unk-nowing to them) were trying to free themselves from the oppressive propaganda coming from all angles in the hegemonic country into which herb was born. The medicine was a funny leaf plant that grew wild in this part of the planet, and not surprisingly, it was against the *law* at this time to smoke this plant. Later in life herb would come to k-now one of the reasons this plant was illegal... because at this time the hegemonic government of the country into which herb was born could make vast sums of black market monies by controlling and running the drugs business, and also make vast sums of monies by arresting and incarcerating large numbers of citizens who were using this plant as medicine, and for recreation, but otherwise doing no harm to others. When a person partook of this plant, they could obtain a sense of calm, and sometimes, providence willing, some insight into that which had heretofore been suppressed by the many nefarious propaganda and pharmacological techniques employed by the government of the hegemonic country into which herb was born. As became well-known in later years, the government of the hegemonic country into which herb was born was involved at this time in many electrical techniques, food poisoning techniques, and medical poisoning techniques whose purpose was to confuse the mind and brain of the population, and this plant helped those who partook to sometimes provide remedy for all these aforementioned nefarious techniques of attack against their person.

On one particular evening after taking some medicine with a group of friends and then returning home to his room, herb was overcome with intense feelings of sadness. herb was actually divining a glimpse into the future, and the vision showed a large separation between, as they say, the haves and the have-nots; meaning that many many future souls would be welcomed onto the planet into a situation where only a very few would have proper sustenance for life, while the others would fight and claw for survival. The tears ran like a flowing river for an hour of herb's time, while he wrote his first essay on the human condition - as he k-new it at that time.

faster Than any animal can run

herb participated in many different sporting endeavors. herb participated in track and field events, and also played soccer (a.k.a. futbol - the beautiful game). On one weekend, herb and friends organized a *Neglected Sports Banquet*, which was held at a park at a local lake. In these days, young people had access to *cars* - a mechanical and electrical machine that one could *drive* wherever they wanted to go. Thus, at this time, a car did not have to register the planned route, and / or gain computerized approval and guidance to get from one place to another.

Also at this time, cars used a combustible fluid called gasoline to provide the energy to move the car from one place to another. The control of gasoline and oil (the raw material from which gasoline is obtained) at this time was very crucial to the sons of Belial (those sOb's). The sOb's made sure that they were always in control of a majority of the planets oil reserves, and thus in a better position to control humanity. This was accomplished in part with endless wars in and around the areas of the planet where many of the known oil reserves were located. As with all the historic battles between the Divine and the darkness, the hegemonic country into which herb was born said they were bringing the Divine (and democracy - sad face goes here) when they invaded various countries on a whim; when in fact they of course were spreading their darkness, and pursuing their world around hegemonic desires.

herb worked one summer in the oil and natural gas fields near the fabled Four Corners area of the hegemonic country into which herb was born. In this area of the country herb came into contact with a tribe of the so-called Native Americans. This was a term for a tribe which was splintered into many sub-tribes around the great land mass that was at this time the hegemonic country into which herb was born. herb did not know the extent at this time of how badly the Native Americans had been treated when future colonists invaded the country 300 and 400 years prior. Over the intervening times, the tribe(s) indigenous to this great land mass were systematically destroyed, or put into FEMA-type camps for *reconditioning*. The sOb's tried their hardest to make sure any real knowledge and the life-style of these Native Americans were banished to the unknown history of these lands. This treatment of Native Americans, as well documented, was how the sOb's had been treating any and all indigenous tribes (whose legacies and legends go back before one of the great cataclysms) world around[5].

the Agricultural and mechanical college

herb left that lower level of schooling and was off to university. He decided that he would study the new area of *computer science*. Computer science was the study of all aspects of computers, from how they are built and organized, to how to

---

[5] so that the sOb's could try to keep sekret much of the real history of the planet

program and interact with them. In these days, a student of computer science still had to take a rigorous mathematics background, which of course suited herb just fine. While herb thoroughly enjoyed his first computing courses, dealing with the details of the ancient and wondrous IBM 360/370 series of computers, he also enjoyed tremendously his *calculus* courses.

In the calculus, a branch of mathematics said to be co-discovered by the shaman Newton and the great shaman Leibniz, a key concept was that of a *derivative*. A derivative, in a mathematical sense, measures a rate of change. For example, if the temperature in a room varies, we can take a derivative of the temperature field with-respect-to the spatial variations within the room. Any non-zero values of this derivative would indicate that the temperature is changing - and thus has a different value from one location to another (but only in the regions where the derivative is non-zero). Thus, since a derivative measures change, if there is not a change in the element values on which a derivative is being taken, the value of the derivative will be zero.

In the particular case of a scalar temperature field, the resulting derivative calculation produce a vector-valued field (aka a *gradient* in this case), indicating both the magnitude and direction of the various temperature changes throughout the room. Vectors were a very important concept in the mathematics as herb was taught. (Vectors, again, are mathematical elements that indicate both a magnitude and a direction; and thus are very different from a simple number - which is also called a *scalar*.) In the calculus, there is a large body of knowledge about how to do calculus with vector-fields, and later in herb's education he would use the so-called vector calculus extensively in his intensive study of *electromagnetic theory* - the general science concerned with that mysterious fluid type magical substance called electricity.

It was not long however before herb decided to study more than just computer science, because as you see, herb was also interested in getting answers to the question: "What is Electricity?"

(i.e., that mysterious fluid type magical substance called electricity). It was thus required that herb change over to an area of study called *electrical engineering*, while at the same time continuing to study the computer science. Sadly for herb, as he so thought at the time, he did never find a direct answer in university to what the true nature of electricity was, but he did come to find out about many of the techniques used to control and analyze electrical activity. At this time recall, the common "types" of electricity were direct current electricity (i.e., the kind you can get from batteries), and alternating current electricity (i.e., the kind that was distributed from many a power plant and delivered to wall sockets throughout the world at these times). For herb, the most interesting type of the mysterious fluid type magical substance called electricity was the alternating current type. And for this alternating current type, to the best herb could discern, the standard way to obtain (or harness) the alternating current type electricity was to *summon* it from the aether of space[6] using rotating magnetic techniques. It was common at this time to setup rotating magnetic equipment which was powered by either running water sources (or damned water sources), or by boiling water and forcing the steam from the boiling water to turn the rotating magnetic equipment. It is from within the organization of the rotating magnetic equipment whence the alternating current type electricity is summoned (from the aether of space). It was also the case in this days that while the direct current type electricity could be stored for later use using batteries, there was no such storage technique for the alternating current type, and thus the alternating current type was simply put onto the electrical transmission systems of the time and thereupon the electrical energy calculations would show that only 30-40 percent of the electrical energy produced (via the aforementioned summoning techniques) would every actually be consumed by electrical machinery plugged into the wall sockets of the consumers throughout the world.

---

[6] there was much contention in those days concerning the reality of an *aether* - a type of substrate that provided foundation for physical events. The great shaman Maxwell most surely insisted on the reality of an aether, but one need not think too hard to figure what the sons of Belial (those sOb's) thought about that.

the World's friendly genius

At this time herb often found himself wondering the shelves of many a used book store. Though herb was being "taught" many things about human (and you k-now who you are) mathematics and science of the day, he had some deep down uneasiness that he was not being told the whole story. Many of the books that herb would stop to peruse were philosophical ramblings from ages past, for example, writings of the great shaman Plato, and ancient so-called religious books that did not make it into the accepted bible(s)[7] of the times. It was on one of these days roaming the shelves with his friend and colleague, the shaman Bonovinci, that they happened upon a treatise written by the great shaman R. Buckminster Fuller, and both herb's and the shaman Bonovinci's life trajectories would be changed forever.

Buckminster Fuller, also affectionately known as *Bucky*, was a Scientist-Artist-Philosopher (sAp) of the highest order. Although he was kicked out of Harvard before enlisting in the Navy in World War I, before his death he was to be presented with over 40 honorary doctorate degrees from institutions of higher education around the world. To understand Bucky, herb always suggested that one should listen to him on tape or video explaining about an early childhood experience building structures in a classroom with peas and toothpicks. As he explains, the other kids "naturally" began building square and cube shaped constructions because those are the shapes of most of the human-constructed things they found in their surroundings. However, for Bucky, because of his extremely poor eyesight, he was not prejudiced by the man-made rectangular bias, so he "naturally" constructed a(n) (equilateral) triangle with three peas as the vertexes and three toothpicks as the sides - as this was a structurally stable two-dimensional shape; whereas a square (made of four peas and four toothpicks) in not so structurally stable. For the rest of his life, Bucky would bravely eschew the normative biases that plagued his contemporaries (and which continue to plague our young today - mostly through mind-

---

[7] 'bible' refers to any *accepted* religious texts of the accepted / organized religions

kontrol programming and political correctness standards espoused by the sons of Belial, i.e., those sOb's).

Bucky's first book was titled "Nine Chains to the Moon," a title that came from the fact that at this time, if the number of humans on the planet were placed head to toe, one after another, then the distance covered by this so-constructed chain would form nine lengths of humans to the moon (and back(2), to the moon again(3), ...). herb would eventually collect almost every piece of information written by Bucky, including the famous Synergetics Dictionary which was a compilation of the cue cards that Bucky and the shaman E.J. Applewhite used to put together the great shaman R. Buckminster Fuller's masterpieces of obscure (at best) Synergetics I/II: Explorations in the Geometry of Thinking[8]. And while Buckminster Fuller is most remembered for his construction of the *geodesic dome*[9], it was the geometry basics that the great shaman Bucky claimed to form the geometry of nature that really caught herb's attention. Before he left the Agricultural and mechanical college, herb would comment to more than one individual that he aspired to build a computer based around the geometrical musings of the great shaman.

the Geometry of nature

Take an equilateral triangle[10] and place it on a table. Add three more equilateral triangles, so that the three added triangles will each share a (different) single side with the first original triangle. Then, bring these three added triangles up from the table (while maintaining the common side with the original triangle - still on the table) so that the three meet above the center of mass of the original triangle, and thus is formed a *tetrahedron*. herb found

---
[8] see the shaman E.J. Applewhite's book: *Cosmic Fishing: An account of writing Synergetics with Buckminster Fuller* for more on this story
[9] see the shaman Hugh Kenner's book: *geodesic math and how to use it* for a masterpiece of presentation of the mathematics of these domes; wherein he writes at the end of the 'What This Book Is' section: "Everyone in this list is indebted – as am I – to Buckminster Fuller, instigator nonpareil."
[10] a triangle with three equal length sides

that the geometry of nature, as the great shaman Bucky described it, used the *tetrahedron* (tet) as the fundamental space-enclosing geometrical form. (the great shaman Bucky would ramble on about how the tet was structurally stable, whereas the basic cube was not structurally stable.)

herb was enthralled when the great shaman Bucky described how to then create an *octahedron*[11] through manipulation of a tet. If one identifies the mid-points of each of the six (6) edges of the tet, and then connects the mid-points across each of the four (4) triangular faces of the tet, we have an octahedron (octa) outlined (inside the tet)[12].

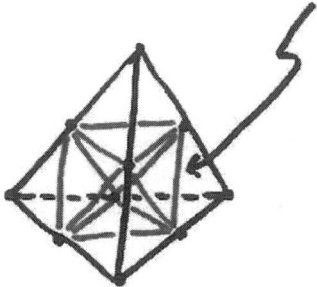

One can repeat this process by identifying the mid-pints of the twelve (12) edges of the octa to find embedded in the octa a *vector equilibrium*[13] (cuboctahedron as named by coexter). This vector equilibrium (VE) was the basis of how the / this physical Universe operated according to the great shaman Bucky.

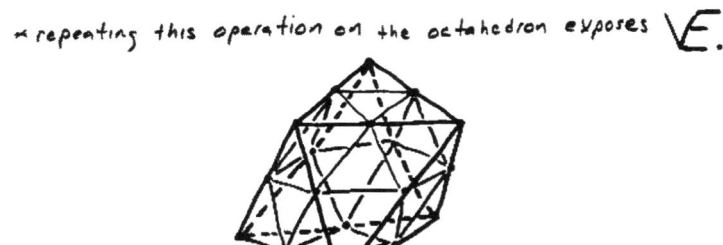

repeating this operation on the octahedron exposes VE.

---

[11] an octahedron has eight triangular faces - compared to the four triangular faces of the tet, and has twelve (12) edges - compared to the six (6) edges of the tet
[12] as shown in the figure
[13] as shown in the figure

Later in his work with the synergetic geometry of the great shaman Bucky, herb would come to understand that one can take the 'vectors of the tetrahedron' (the six edges in a proper spatial arrangement) and move the three vertices of the original triangle (pictured still lying on the table), each along with two connected edges of the tet (in a proper spatial arrangement) to the center of mass of the original triangle - which becomes the center of the VE - so that these six edges of the tet - the 'vectors of the tetrahedron' - now connect the center of the VE with six of the twelve outer vertices of the VE[14]. So that now, starting with a tet, we find four levels deep, the tet again... which would please the great shaman Arthur Young, whom showed that after four (4) 'derivatives' (operations) we should get back to what we started with (as is what happens with the sine and cosine functions).

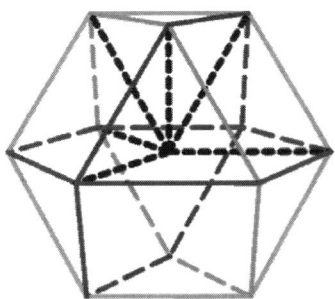

old Places

herb was old enough now to contemplate some of the stories told to him by one of his uncles, the great shaman Mason. Mason was a controls engineer, who happened to be one of the top engineers in the world at this time for designing the controls for that mysterious fluid type magical substance called electricity;

---
[14] as show in the figure

especially when it came to very large power projects (associated with dams, airports, etc.). As such, the great shaman Mason had occasion to travel around the world to work on projects, where he would spend many moons installing the controls.

The great shaman Mason would describe to herb his many adventures in these far away lands where he would investigate in detail along with Dorothy the various archaeological digs he would happen to come upon. It always fascinated the great shaman that some of these excavations would uncover 6 to 8 (or more) different periods of civilization, each built on top of the others. herb and the great shaman Mason would speculate on how far in the past some of these earlier civilizations lived, and they agreed it could have been up to 1000 or 2000 years (or so) between each – realizing of course that one had been covered up (literally) before the other was built above it. Thus, some of these dig sites recorded human (and you k-now who you are) history 10-, 20-, maybe even 50,000 years (or more) into the past. It made one wonder about the accepted history of the time of course, and to whether it was true that Mesopotamia and Eqypt were in fact the first real civilizations.

Later in life herb would study the writings of the great shaman Michael Tsarion and the great shaman Immanuel Velikousky, who both had suppositions on historical events in the long ago times on our planet that were not being taught to the school child at this time. Again, as herb came to k-now, here was yet another example of how the sons of Belial (those sOb's) did not want the people of this 'modern age' to k-now their true history. Evidently - as herb would come to k-now, the sOb's wanted / needed their carefully constructed (false) history to be the accepted history – and anything in the past from these / those old Places were / are summarily destroyed, ridiculed, and / or genocided.

The great shaman Mason would also discuss his fascination with the idea and realities associated with hypnosis. The great shaman would amaze herb with stories about how a hypnotized patient / volunteer could, for example, be made to lie between two chairs and settle in to be as stiff as a door, and support, for example, the weight of four others. A hypnotized patient / volunteer could have their leg or arm punctured with sharp instruments without notice, or (as the sOb's k-now well) be made to remember things that never happened, or forget things that did happen.

old Times

When the current physical incarnation of the great shaman Mason's immortal soul passed from the / this physical realm (may the Divine bless his soul), herb was given early access to pick as many books as he wanted from the great shaman's library. While herb selected many of the works by the great shaman Immanuel Velikousky, there were other books that also presented suppositions, about not only old Places, but hypothetical descriptions of the associated old Times.

Later in life herb would happen upon related information that purported to show that most of the accepted ancient religious texts / control manuals were in fact fabrications that plagiarized, or blatantly stole previous myths / legends / stories from other cultures (associated with old Places and old Times). The early six books of the chosen people, for example, was possibly / probably put together in (the ancient learning center of) Alexandria, Egypt from available material in the famous library. It could be the case, as herb came to understand, that the whole story of these chosen people was most likely fabricated by an ancient Arabian tribe known in earlier times as the Hyskos. After infiltrating a somewhat serene open culture in Egypt - which did not k-now slavery at that time, the sons of Belial (those sOb's) took over the throne in the form of the demigod Akenatum (the first sorcerer?). This demigod under no uncertain terms trashed / vandalized / corrupted the teachings of the blessed land. Among other things in the apparently false history of the chosen people, the so-called exodus was a flipped-upside-down account of when this nasty-to-others tribe were thrown out, and banished from the lands of Egypt, where / when in fact even the foreign armies the sOb's had brought in to Egypt, could no longer protect them from this rightful / justifiable banishment. It may have been the case, that the invader Venus was also making trouble in those days, the stories of which are also most certainly corrupted in the false control manual(s) of the sOb's.

herb would come to think that the oldest of old times however came to us both in stories flipped-upside-down in the books of the sOb's, and from the great shaman Plato. The great shaman Plato mentions here and there in a couple of works - albeit slightly romanticized - some of the exploits of the fabled island civilization of antiquity whose capital was Atlantis. The Atlantian myth / story / accounts date into antiquity, and it could be the case that participants in the battles between the Divine and the dark from those days continue their battle today. herb came later to learn that the Atlantians fell from grace in no small part due to their complete lack of civility that eventually developed. That is to say, their ruling class may have became first class sorcerers, the first troupe of these ancient sons of Belial, and were (and apparently remain) hell bent on reclaiming the entire planet, as it pertains to both physical control, and the control / direction of spiritual matters.

17 Years in the wilderness

Over the next 17 years herb undertook many adventures, and at the same time became somewhat acclimated to the world as it was (or as it appeared to be... which may or may not be the same thing?). herb worked as an engineer for a large international corporation for a while, but eventually got very bored with the job as it became mostly a matter of keeping track of the monies instead of any work that was actually technical or interesting. herb dabbled in some graduate courses in mathematics and computer science while working, and then eventually decided to go back to school full time.

herb spent 7 full years in graduate school and earned three graduate degrees, including the PhD. During these years herb began to realize that there was truly not a mathematics that he could not understand when he spent the proper amount of time studying the publicly available information on any particular topic (in mathematics). herb was particularly fond of the mathematics associated with electromagnetic theory (that is, the mathematics that described how man had come to harness, distribute, and manipulate that mysterious fluid type magical substance called electricity).

herb also continued in these days to work with and study the synergetic geometry of Buckminster Fuller, and before he left university training, he clandestinely submitted a paper proposing a ground-level theory for how the Universe computes, an area of study that herb coined *computational Cosmography*. herb also secretly was able to encode the electromagnetic theory (of the electric and magnetic fields) onto the synergetic geometry.

herb did make an early effort to patent his new technique for understanding and computing the relationship between the synergetic geometry of nature and that mysterious fluid type magical substance called electricity, but, as he would learn later, the sons of Belial (those sOb's) would just as soon use someone else's inventions for free (in their sekret, deep in the shadows, quasi-government labs) than pay fair compensation. herb would often think later that he was a bit lucky not to get a patent at that early time, as it was standard practice in those days for the sOb's to coerce information and code from inventors and scientists, or lock them away on fabricated charges if they did not play ball by their rules. The same of course routinely happened in the fabled land of Holly-wood, where many an aspiring writer would have her / his screenplay stolen and used without permission and / or without fair compensation.

herb spent his last year in the wilderness in the high desert of the Amerika southwest (in the hegemonic country of his birth) surrounded by mountain peaks in a valley that contained the head waters of the Grande River. herb had a job there as a university teacher and he liked it very much. herb was able to help others learn mathematics and computing techniques, and while he made sure to offer a real challenge to the students, herb also tried to be reasonable in his expectations.

Before herb left the high desert valley and come forth out of his wilderness, he ended this sojourn as he had begun it 17 years earlier, with a visit with the spirit of the blue ring - as shaman and mystics had been doing from time immemorial. This was also of course one of the final secrets of all mystery schools, in that they prepare the initiate for a special psychedelic experience of one kind or another, after which it is easier to indoctrinate the initiate into the particular flavor of high-weirdness understanding that any particular school was advocating.

into The new world order

herb came out of the wilderness and began an apprenticed professorship towards the later part of the Gregorian calender so-designated year of 2001. This was that same year the now most infamous of false Flag events occurred, the 911 event (which also at that time was the standard emergency number in use in the Amerika phone system[15]). It has, of course, in our time been shown without any doubts that the 911 false Flag event was well and truly orchestrated by what herb would come to call the *cabal of maniac magicians* (i.e., the sons of Belial - those sOb's).

You see, even though many issues would come out right away about how this event was not really what it seemed, the mind-kontrol capability and techniques of the sOb's was so pervasive, it

---

[15] and as it was, this annual date on the calendar was the anniversary date for the forming of the Pentagon – that enclave of demented war pigs, traditionally at the service of the sons of Belial (those sOb's), until that final insurrection of Divine sanity finally overcame the darkness so loved by the cabal of maniac magicians.

would be another 10 years until herb himself would come to know anything different than the "official story." As you might surmise, herb cried and cried when he came to understand the depths of depravity to which this cabal of maniac magicians (those sOb's) had stooped, butt at the same time herb openingly admitted that one had to admire the commitment and simple work ethic that the sOb's and their minions (some in the k-now, and some not) had shown over the many years (OK, centuries!) in order to pull off such a plot - along with completed storylines, draconian legislation at the ready, and some other worldly money-making from the market on the side.

During herb's apprentice professorship training he concentrated on how best to get students to study and learn. herb was interested in getting students to *think* deeply about the technical areas he was covering, and not so much interested in simple rote or parroted learning. Now you can imagine how this got herb into various bouts of trouble in the university systems of that age, where the sOb's subversive activity over so many years had created an environment wherein university no longer had anything to do with helping students on a path to being able to think, as they (the sOb's) wanted above all else, subservient compliant worker slaves for a planned technocratic new World order (nWo).

As it aforementionedly happened, it would then be another ten years before herb would be required, by providence, to revisit any consideration whatsoever of what was then so appropriately designated as *conspiracy theory*. You see of course can you not, that this type of label, which uses a valid word such as *conspiracy* - a secret plan between conspirators - can be turned into a label for ridicule (or a badge of honour?) by the sOb's. As it was, in these earlier years, herb's attention was required elsewhere, lest his preparation and further training be wasted on a life of luxury, greed, hubris, fear and hate - as was of course expected (and respected - sad face goes here ;-( ) of a proper citizen of the hegemonic country of his birth in those days.

are You living in a computer simulation?

herb at this time was further developing his geometric, electromagnetic, and philosophical computation ideas. While he prepared more papers for publication, his family was blessed with the companionship of two new embodied souls into this physical realm of the perpetual now, which brought the family total to 5. herb at this time started to pay attention to the then rapid rise of the police state, and herb took the decision to move out of the hegemonic country of his birth, and head towards a fabled land where they spoke the old language, and where one can find a sun that never sets (during a certain part of an Earth year).

It was as herb was preparing his move out of the hegemonic country of his birth that he first came across the shaman Nick Bostrom's seminal paper "Are you Living in a Computer Simulation?". herb had an affinity for understanding all aspects of the so-called simulation argument[16] presented in this paper. And while herb was uneasy about parts of the argument (for reasons not yet articulable by herb), he spent most of the entire summer of the Gregorian calendar so-designated year 2004 studying / discussing / re-reading this paper written by the shaman Bostrom.

---

[16] In the simulation argument, the shaman Bostrom supposed there were three possible situations we humans (and you k-now who you are) at the time found ourselves in. As will be explained, the shaman Bostrom also thought, via his highly above average reasoning toolkit, that the three situations were also disjoint - meaning that only one of them could be true. Thus he referred to them as disjuncts. The first disjunct stated that humans (and you k-now who you are) on our planet and any/many other planets where they may reside would most likely never reach a 'posthuman' stage. (Posthuman referring to an evolution of sorts such that concerns of war, disease, poverty, et. al. would no longer be a factor.) The shaman Bostrom provided possible reasons why this may be the case, and the juvenile behavior of large groups of humans (and you k-now who you are) on the planet at the time (mostly/partly due to the negative influences of the sons of Belial (those sOb's)) lent credence to this possibility. However, what if some planetary groups of humans (and you k-now who you are) on some planet did reach posthuman stage? This led to the second disjunct wherein the shaman Bostrom claimed that posthumans would decide not to run / create what the shaman Bostrom called *ancestor simulations*. The ancestor simulations of this story (and second disjunct) were presented to be a very realistic type of computer simulation, where - a key point here - where the computational elements he called ancestors were so well programmed, that they (the computational elements called ancestors) 'believed' themselves to be living entities populating a Universe such as ours. So then, to recap, either (1) human (and you k-now who you are) societies never reach posthuman stage, or (2) posthuman societies never build and run ancestor simulations. (Pick one or none.) If both of these first two disjuncts are false, then the shaman Bostrom supposed (through statistical / probabilistic / large-number reasoning using that above average reasoning toolkit of his) that *we must be living in a computer simulation*. Now one can see how the shaman Bostrom came to this very appropriate name for the paper where this argument was presented

herb was attracted to this paper first of all from his computational leanings that he had articulated in his computational Cosmography, wherein (wherefrom) his vision included all phenomena (in physical Universe) utilizing an energetically dancing basement of geometrical forms. Thus, herb had a belief (as did the great shaman Buckminster Fuller) that geometry underlies all physical and metaphysical phenomena. Because, as you may come to understand, one of the disjuncts of the shaman Bostrom's simulation argument was that the Universe was a simulation. (Albeit a naive simulation as described by the shaman Bostrom, based only on a digitally encoded information theoretical amount of content - much like a quantum theorist of those long gone days who surmised that Universe computation was only about the bits, sans concern for the underlying form and the required interactions with a non-physical - and possibly the divine.) Another part of the simulation argument that really intrigued herb was the second disjunct, wherein the shaman Bostrom posits that the fraction of posthuman civilizations that would run *ancestor simulations* is very close to zero. This troubled herb for reasons unknown at the time, but the desire to come to terms with his distaste of the disjunct would lead herb onto a path wherein he would have to come to terms with the dualist nature of mind and matter (in the great shaman Descartes terminology) - eventually leading herb into the needed study and understanding of theurgy.

what Is a computer?

In reading the simulation argument as presented by the shaman Bostrom, it came to make perfect sense to herb that the Universe we appear to be embedded in is in fact some type of (computer?) simulation. However, herb realized that this begged the question as to what Is a computer (and what Is computation).

herb understood computation as it was k-nown at the time very deeply, and he understood that the ground-level of computation involved the control of the flow of that mysterious fluid type magical substance called electricity. It was then the human (and you k-now who you are) interpretation of this coordinated control of that mysterious fluid type magical substance that resulted in a meaningful computation. herb also understood, that with the proper abstraction, one is able to veiw computation simply as the manipulation and control of the flow of number; where, for example, this idea of number flow can result (in one sense) is we abstract up from the flow of that mysterious fluid type magical substance called electricity and focus on all the *bits*[17] of the computation.[18]

So then, if the computer was simply an orchestrated dance (of sorts) involving flow of that mysterious fluid type magical substance called electricity, or an orchestrated dance (of sorts) involving flow of number (if one employed that abstraction), then it is not much of a leap to consider the whole Universe as a type of computation, wherein, herb reckoned at first blush, it could be that fabled quantity of *energy* that makes up the flow that is controlled in the Universe (as computation scenario).

the Heresy of scientific materialism

During his intense study of the simulation argument, herb was now beginning to understand his uneasiness about the particulars. It was the second disjunct of the simulation argument that troubled herb most. Here is where the shaman Bostrom supposed that (in some distant future time, or some distant past time, or some different place) posthuman scientists made the choice to run *ancestor simulations*, wherein, one may suppose, the shaman Bostrom envisions these type of (computer?) simulations are capable of

---

[17] that flowed to and fro along the wires, or within the crystals of the computers of that time.
[18] The great shaman Pythagoras claimed that "things are numbers," and that mathematical formulas and ratios explain the physical world. Pythagoras, following Thales, spoke of the world as a cosmos, or an order, indicating that order was the essence of the Universe; that laws, number, proportion, or symmetry was the universal principle of all things. Now, for the Egyptians, numbers were the most ancient form of symbols, and represented the energetic formative principles of nature; and they were called 'Neters', which has its hieroglyph translated as 'gods', according to the shaman de Lubicz

capturing the feelings, desires, etc. of the humans (and you k-now who you are) that are being simulated.[19]

This is where herb had to take a stand on the nature of mind and matter, and herb found that he was not willing at this time to presume that all mental activities that a human (and you k-now who you are) experiences is simply the result of physical phenomena interactions in the brain. You see, that in these times, some / most scientists had come to the conclusion that the Universe was simply a conglomerate of energy events, and the belief among many was that all such physical events could be described in informational terms, and as a result, the information pertaining to all such events could be stored on a computer. Such scientists, herb came to understand, were properly called 'materialists.' herb also found that materialists who believed their techniques could also be applied to mental activity (and everything else as it concerns brain / mind) were called functionalists, or sometimes computationalists. (herb would later call a certain subset of these functionalists the 'Brain builders," and would often cite the 'Brain builders dream.')

physical Universe needs a non-physical

A key problem you see is that herb was familiar with the quantum mechanical theories of the time that required what was k-nown as the quantum vacuum mechanism. The quantum vacuum supposed that for *any* particle interactions (e.g., any interactions involving electrons), there was required a large flux of antiparticles and virtual quanta that literally popped into and out of existence (as so eloquently described by the shaman Wheeler). And this flux of 'stuff' had to come from somewhere / someplace / some-abode - or so herb reckoned. The shaman Dirac made reference to an 'infinite sea of positive electrons," which then could be accessed on the whim and as needed by all physical process[20].

---

[19] The purpose of such simulations herb could not quite discern from the shaman Bostrom's descriptions. Suffice it to say that such ancestor simulations might be targeted in order to reveal a more realistic historical narrative concerning how the posthumans arrived in their current predicament(s).

[20] within the participation limits specified in the Heisenberg Uncertainty Principle.

The consequence of having a quantum vacuum mechanism to support all physical process "in Universe," made it clear to herb that other 'things' / 'stuff' besides just the physical stuff (i.e., energy events) was required if one was to create an accurate Universe (as we k-now it) simulation. Thus, herb was convinced that the shaman Bostrom's supposition of simulating mental activity was not going to work as the shaman Bostrom supposed.

Later in life herb would apply this same analysis to the suppositions of various representatives of the sons of Belial (those sOb's) who envisioned placing consciousness inside a machine of man's creation. There were those who dreamnt / forecasted and / or prepared for the so-called *singularity*, that they (i.e., the sOb's) would bring forth from their deep dark double top-sekret labs - whence the Brain builders dream.

what Descartes supposed

herb kept on studying the issues associated with simulating a Universe (as we k-now it) beyond the simple type of masquerade proposed in the ancestor simulations of the shaman Bostrom. It was during this study that herb rediscovered some much earlier work by the shaman Penrose (whose book: *The Emperor's New Mind*, herb used to have in his possession).

The shaman Penrose was a scientist of the highest caliber who evidently was able to escape some (but not all) of his materialist conditioning. The shaman Penrose came to the conclusion that mental activity could not solely be a result of physical interactions (involving energy events in physical Universe), and that something else was required[21]. Now, herb did eventually also come to understand that the shaman Penrose did not take his notions to the same conclusions as the shaman Descartes had some 400 years or

---

[21] The shaman Penrose even gave a nice 'proof' of this fact, which reminded herb of the famous logical result brought forth by the great shaman Godel, who showed that every logical system has some truths that can not be proved within the logical system... thus any logical system can not be closed, and something else was needed. The great shaman Godel's work here was the reason many of the workers on the more modern Principia Mathematica project abandoned ship; as they had a result that told them they could not succeed.

so earlier, but it was a nice start. You see, the shaman Descartes clearly stated that there was matter stuff, and that there was mind stuff, and these two stuff-ings were of a different nature. The *res cogitans* (mind stuff) also need allow for non-temporal, non-spatial activities; a requirement that would be rediscovered by the great shaman David Bohm when he grappled with the meaning of quantum mechanics and the mind-matter relationship.

The shaman Penrose eventually supposed that deep down in the cubby-holes of space-time, there must exist that extra 'stuff' needed for consciousness, and interestingly named the abode containing all these cubbies a *Platonic World*. It was then via access to this Platonic World that the extra stuff (beyond the physical) was accessed in order for, say, minds to ramble on with their thoughts, intentions, desires, et. al. - or so - the shaman Penrose proposed.

let Me get it back

It was at this point in herb's studies that he recalled his many hours with many books related to an older mystery religion idea that in these days was called Gnosticism. herb would later surmise that even though there were attempts to eradicate all evidence of this ancient mystery religion over the many years by the sons of Belial (those sOb's), it was likely divine providence that the ancient Dead Sea scrolls and Nad Hammadi library were unearthed so that this ancient way of thinking was available for herb's and others perusal. The Gnostic system, herb realized, was germane to his current studies of the simulation argument precisely because the Gnostics believed that this physical realm was actually built by a so-called *Demiurge* (a lower-type, a want-to-be-, or even maybe, a dark-god, if you will). Thus, the Gnostics thought of the Universe as a simulation (of sorts).

The Gnostics also held to the tenet of the immortality of the soul, and espoused the belief that it was (possibly?) a mistake for the Demiurge to build this / the (lower) physical realm, and whence entrap within 'it' the divine spirits / souls that had a proper home in higher realms / abodes. Thus, the Gnostics held a belief similar to Cartesian Dualism, in that the mind / soul and matter are of different natures.

baby Where i come from

This Gnostic reckoning that an immortal soul is entrapped in this / the physical realm danced very nicely with the problems associated with the Heresy of scientific materialism that herb was wrestling with at the time. herb would many times step back and compare the quantum mechanical theories with both the Cartesian Dualism (of mind and matter) and the ancient Gnostic reasonings (which the sons of Belial (those sOb's) did not want anyone of these times thinking about).

It was around this time that herb made the decision to view the quantum mechanical theories not as the 'real stuff' of Universe, but only as a description of some of the machinery involved in the presenting of this elaborate / beautiful / unexplainable(?) play / drama that some called the Universe. With this new framework on viewing Universe, herb revisited his geometrical theory of how the Universe works, and began work on updating the computational Cosmography.

computational Cosmography and all that that entails

computational Cosmography supposed to use the geometry of Nature (as supposed by the great shaman Bucky) to outline how the Universe – in total – computes. And while herb in time would realize the idea is of course totally absurd (that is, to suppose that humans (and you k-now who you are) could really ever totally understand 'reality'), the exercise of thinking of the Universe as a geometrical-type computation seemed worth pursueing, because, after all, it was well k-nown in these times that mathematics could be used in the engineering of machines and computing devices of the utmost complexity.

In the computational Cosmography as herb came to understand it, Nature employed the VE (vector equilibrium located about the vectors of the tetrahedron as shown in a previous figure) model and the synergetic geometry internals thereof for both the framework of 'things' in Universe and for the entity called 'space' in the physics. To wit, herb followed the great shaman Bucky's idea of combining many VE together by making the exterior vertices of each VE the center of its own VE (and so on) to build up what the great shaman Bucky called the isotropic vector matrix[22] (IVM). The figure that herb would use to explain this point shows two 'Mite's (made of 2 A-modules[23] and one B-module[24]) within an IVM[25].

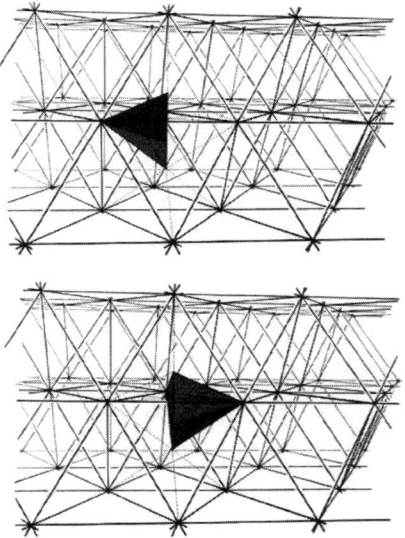

---

[22] which is also called a face-centered cubic (FCC) structure; which happened to be used to organize the majority of atomic solids.
[23] an A-module, for example, is one-24th of a tetrahedron, obtained in a beautiful geometrical decomposition of tetrahedron, in a typical fashion as many other geometrical presentations in the great shaman Bucky's synergetic geometry explorations.
[24] to make a B-module, the geometric decomposition is done on an octahedron.
[25] here a copy of one of the color plates from Bucky's Synergetics books.

Now, this is just a static *(stay)* picture of some geometrical forms that herb used to start to explain the computational Cosmography. herb would then typically try to get someone to envision this as part of 'What is,' by imaging a dynamical flow of many forms interacting within a background lattice that is also evolving dynamically. In a paper herb presented in Amsterdam (cough, cough), herb had this to say about the computational cosmographical interplay of space and matter:

> Both the background lattice system and the big bits (or polyhedra aggregates) of a CC computation have restrictions on form resulting from identical embryonic geometrical considerations. It is from the omni-directional lattice system that the form for all entities of a CC computation emerge. The interplay between the big bits and the lattice system creates the need for an active lattice background system, and was described in (a paper presented at the Santa Fe Institute) thusly:
>
>> In this way, we find the lattice tells polyhedra aggregates how to transform, while the polyhedra aggregates in turn tell the lattice how to tense

herb conjectured, as did the great shaman Bucky, that it was the tension, or 'tensegrity' (a Snelson term adopted by Bucky) of the background lattice system that produced the phenomena that was called 'gravity.'

Now, to use this 'toy model' as a way to think about Universe (as computation), herb added many caveats. The main caveat is how to think about the quantum vacuum mechanisms, wherein it is required that anti- and virtual- particles pass into and out of existence (in physical Universe) as part of any physical process.

This *reality flux* mechanism must then both power a 'jitterbug'ing of the background 'space' lattice[26] and allow a communication channel (of sorts) between physical Universe and, what herb came to call, *Negative Universe* - the abode of, for example[27], 'things' of the reality flux when the 'things' are not in physical Universe.

herb, many a time, especially in the early days of presenting ideas associated with the computational Cosmography, would invite the audience to visualize large scale interactions[28] of the geometric forms, and then consider the wonder of scale associated with a dance such as this that would go all the way down to the Planck scale of the quantum foam, up to the 'particles' that make up the Standard Model, then to the atomics, then to the chemical / molecular level, then to the biologicals, then to the creatures (including humans - and you k-now who you are), then to the galaxies, and upward to, what, our Hubble can resolve. Typically, it was the artists in the audience who would later explain to herb about the 'goose bumps' that accompanied their envisioning exercise herb had tasked for them.

It was after this level of understanding that herb also tasked himself with the wondering about how the mental world would interact with such a dance of (geometric) forms and 'space' so constructed. herb had no choice he realized, but to take into account all of the Gnostic, Cartesian, and reality flux reckonings, and take hints from the quantum consciousness studies of the shaman Penrose and the shaman Hameroff. Thus, herb was in agreement with the great shaman Bucky, and thus supposed that mental activity 'resided' in the Negative Universe. Further, herb realized, this abode that herb called Negative Universe was connected to all of physical Universe simultaneously via the 'reality flux' mechanism; and that this simple realization then captured (in an

---

[26] thus creating a need for both virtual VE as part of the 'space' lattice, and the expansion / creation of space

[27] many other items particular to what we call Universe may also be part of this abode, including, the morphogenesis form of 'things' that would be said to possess 'life' in physical Universe, and the mental forms that make up the Minds of wo/man.

[28] really dances!

Ockham's razor sort of way) a mechanism for explaining quantum non-locality, the great shaman Plato's 'Divine Mind,' and the shaman Huxley's 'Mind-at-Large' ... and many other things that herb supposed lurked in the shadows, including his (imagined or not?) arch-enemy-entities the 'archons'[29].

witch Will it be

herb often wondered if there was a time when wo/man did not have the burden of choice, or the burden of dealing with their 'will.' herb did come to understand that this was an important question when one keeps a 'belief' in the immortality of the soul; wherein this soul / spirit / 'connection to the divine' is a type of background 'driver;' butt not so much like the Id of the shaman Freud's wonderings, but more of a way to use a built-in connection to the divine if one will-ed it such.

Now herb k-new there were other animal-type things on the planet at this time that built things, for example, many birds built nests, many part-time water creatures built dam-type things, and many different types of animals and insects built living quarters under the ground. Some animal-type things maybe even built tools ... but herb was not sure of all that.[30] Then, of course, there was that Hindu 'belief' that a soul / spirit must go through many different 'incarnation' experiences on 'its' way to Brahma-hood, some even in animal-type things. So it may be possible that these other creatures, besides humans (and you k-now who you are) had also a soul / spirit incarnation ... so in the end herb admitted to

---

[29] the etymology of this word *archon*, as fable would have it, is traced back to a word that was used to also describe 'petty rulers,' or 'petty administrators', i.e., some individual whom was presumed to rule by fiat, and/or on a whim, and you of course guessed it did you not, these whims would rarely be about concern for the humans (and you k-now who you are).

[30] ancient pagan thinking put spirit-essence into many 'living things,' including not just animal-type things, but plant and tree 'life' as well, and typically gave 'thanks' for all that was around them. AND BTW, most of these type of pagans DID NOT practice sacrifices, as this is one of the oldest pys-ops ever; meaning it is an age ole practice of the sons of Beliel (those sOb's) to turn a story around and accuse their enemies of the nasty things that they (i.e., the sOb's) practiced. Oh, and practice such atrocities they did, and still do, with their (i.e., the sOb's) more modern means of sacrifice including not just war, but rampant kid-napping and murder also.

himself, that maybe a child can help where they are born. herb felt in quite the quagmire as he wrestled with the Objective reality of a computational Cosmography-type Universe which utilized a reality flux as its 'power source,' and, apparently - a sentient being would hope - allowed (demanded?) the incarnation of an immortal soul within each human (and you k-now who you are). Thus, the human (and you k-now who you are) would have choices, call it free will if one wants, but choices wherein each human (and you k-now who you are) could simply do what they were told, or what they were conditioned to do, or they could take control of their 'will' and do what they wilt.

herb of course was aware that in those days many of the apparent choices made by humans (and you k-now who you are) and the sons of Beliel (those sOb's) exhibited what could be called evil intentions, while many other actions of humans (and you k-now who you are), the vast majority maybe, were of a more civil nature when it involved a fellow wo/man. For each human (and you k-now who you are) it seemed they had a choice if they choose, and so witch Will it be as it pertains to each individual's burden, Will it be theurgy or sorcery? Of course herb was also well versed in the fact that many older societies had strict laws against the so-called 'black magick' (or sorcery), and that magic in both flavors, both divine theurgy and dark sorcery had been very widespread amongst older societal populations, as witnessed, for example, by the many 'curse scrolls' (or 'tablets') unearthed in both Babylonian and Roman excavations.

onward Towards a need for theurgy

In many of herb's esoteric studies, involving, for example, not only the Socratic dialogues of the great shaman Plato and old Egyptian writings, but also many Theosophy writings, and many entries in his library by the shaman Manly P. Hall and the shaman Israel Regardie, and even some works by and about the great beast Aleister Crowley[31], herb found reference to encounters with entities / 'spirits' that were not deemed as human (and you k-now who you are). herb by this time was fully aware of terms such as astral planes, ethereal world, Hidden Masters, Askashic records, and so on, and, of course, with the control manuals' use of the terms 'angels' and 'demons'[32]. In herb's Gnostic studies, the term 'archon' would fit as a blanket term for *any* non-corporeal entities witch participated in the machinations of the sOb's, whose likely residence was of the astral / ethereal realms, and these 'archons' would typically be of the inclination to 'tell,' or 'command,' or 'coerce' a human (and you k-now who you are) into any action whatsoever.

As it was however, can not you see, in herb's time in the hegemonic country of his birth, the sons of Beliel (those sOb's) and their corporate media minions, education destroyers, and cultural narrative writers made every effort possible to 'play down,' or discount the experiences[33] each human (and you k-now who you are) had in their 'dream worlds.'[34] However, as herb well k-new, and as did even the peasants of ancient times, this 'dream world' was an important part of 'life.' Butt alas, herb realized that each individual may be able to be convinced, that is, each one of us humans (and you k-now who you are) may be induced by many

---

[31] much of course has been written about the 'great beast,' who was an agent for the sons of Beliel (those sOb's) in both world wars organized by the banksters of the City of London, Butt it should be pointed out, that the 'great beast' did provide a service with his insistence on publishing many of the ritual techniques practiced by the sOb's, thus leaving no possible doubts about the prevalence of this type of Enochian magick brought into our modern times by the dark magus John Dee - the original 007 in her majesties sekret employ.

[32] It was agreed amongst most scholars in these times that 'demon' was a play on the original 'daemon,' which was more like a subordinate deity, or even an individual's 'attendant spirit,' so as to say, 'everyones got one;' which under other mythos was also called an individual's 'guardian angel.'

[33] STOP THE DREAMS was also a mantra of the sOb's, and they had many a chemical and electromagnetic technique that would attempt just that

[34] the dream world of the Australian aborigine may be more 'real' to them than the 'real' physical Universe, and is in a matter of fact way also used as a communications channel amongst the population. This explains does it not why the sOb's had to dish out such brutal treatment to the aborigine in their conquest of this continent; as the sOb's want nothing considered 'real' unless they tell 'us' it is real.

of the reality-deforming techniques used by the sOb sorcerers to 'believe' that the dream world does not 'mean' anything, because, after all, as was well k-nown in those times, these dream worlds are not built of material things, and the sOb's made many efforts to convince that anything non-material is not 'real,' and thus not part of 'real-ity.'[35]

herb thus had no choice from a scientific point of view butt to believe magic[36] to be a 'real' thing, permeating all existence in much the same way as 'light,' from which is built the complete cosmographical physical Universe. herb was adamant that magikal techniques naturally split into two pretty much disjoint categories: theurgy and sorcery. Sorcery, herb contended, involved any techniques witch made efforts to subvert or manage the 'will' of others, typically using techniques not k-nown to the victims (or targets) of the sorcery.

With this k-nowledge of how magick is practiced, herb realized right away that most propaganda and most advertising techniques in those days were both 'straight-up' sorcery, because, can you not see, both propaganda and advertising used subtle techniques to manipulate humans (and you k-now who you are) into 'believing' things (or 'wanting' things) that may not be in the targets best interest. Theurgy, or divine magick, on the other hand, did not use techniques meant to override the 'will' of humans (and you k-now who you are). Many theurgetic techniques are available for divination (i.e., looking into possible futures), personal well-being, and management of one's immortal soul, and these theurgy techniques do not require that the practitioner take advantage of others (and / or subvert the others 'will').

---

[35] except, of course, quite strangely, for that one non-material god in each of the control manuals of those times in the hegemonic country of herb's birth.
[36] for an easy definition of magic one can say that magic is the manifestation of 'things' in accordance to the 'will' of the practitioner. Thus, in a very real way, even the building of things created in ones imagination is magic, as is most art, or anything that results from planning. To wit: all deceptive planning of any group, e.g., the sOb's, who impose statecraft on the general population, these sOb's are performing magic, and since it (i.e., statecraft) usually is against the 'will' of the population upon witch this statecraft is performed, their (the sOb's) magic of statecraft is properly called sorcery.

hmmm ...

herb was now determined to try to explain all of these issues that divine providence had placed before him, all these things that seemed to be associated with the general issue of 'on What is.' For one thing, herb k-new well that one always learns more about what one is thinking about when one makes an effort to explain same to others. Further, from as long ago as herb could remember, he was interested in understanding 'What is,' so why not study it some more.

herb thought it best, after many moons of contemplation, to employ a dialogue technique in his writings on these many subtle subjects, you k-now do you not, the type of dialogue found, for example, in the writings attributed to the great shaman Plato and the great shaman Giordano Bruno. herb thus set out in the end to write up a 'play' - of sorts - containing three acts, which was really more of a religious tome, and he was granted title for the three acts: on Physics, on Mind, on What is

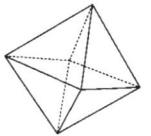

INTER-MISSION

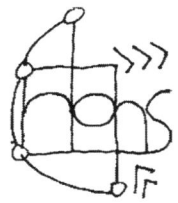

once Upon a path

> *today was such a great day; the weather was so weatherful; imagine just like the Indian summers the Indians must have enjoyed many moons past. It is strange how mountains, valleys and rivers still must hold the same vision for my neighbors and I as it has for those before us. Such beauty and integrity*

thalamus: is not this a nice place to visit?

Glia: i have never been here before, but it is nice. Nothing else around but this lake and rock cliffs; and of course some other people.

thalamus: i did not really notice. (lighting a cigarette) Feel like adventuring tonight?

Glia: what did you have in mind my dear thalamus?

thalamus: just going places that we have never been before, and will not be able to get back to after we leave.

Glia: why do you talk that way, you do not have to joke about everything. I am not sure though; will we get lost?

thalamus: only if we are not careful about the paths we take; in the dark everything is different and new. Even if we do get lost, we probably will not be able to tell for years, or at least until we are not lost anymore.

Glia: i will think about it.

thalamus: about what?

Glia: about adventuring.

thalamus: you mean going places you have never been before?

Glia: that we have never been before, and will not be able to get back to after we leave.

thalamus: what kind of place can one go to and then never return after one has left?

Glia: i do not know; maybe it has to do with the past. People say you can not travel back in time you know.

thalamus: i have heard that said.

Glia: but that really does not hold water like that lake does.

thalamus: maybe that is it.

Glia: you mean like the lake does; like it is really not a place at all.

thalamus: now I understand what you mean. If it is not a place then that explains why you can not return there, wherever that is.

Glia: now you get it. It is more related to a thought or a way of feeling yesterday, which is not today, because it is yesterday, which is not today. Get it?

thalamus: why do you sometimes talk that way? Not that I mind, mind you.

Glia: i will think about it.

thalamus: about what?

Glia: i am not sure really.

thalamus: how about a walk. We can talk about simple things, like why a rock is a rock and not a tree, and why a tree is a tree and not a rock, and why they are not the same thing; in which case we would then be free to talk of other things.

Glia: sounds Beautiful.

> *i think I will take a walk to the river tomorrow. It is so nice this time of year with the leaves changing and all. That always brings back nice memories*

theurgist Of atlantis

> *herb and his band, Theurgist of Atlantis, are up next. They typically play the same songs they sang the week prior, but sometimes an old one is skipped, or a new one pops in. This repetitiveness notwithstanding, the show basically goes as follows...*

herb: mystic Tomato, mystic tomato *(in a highly whiny, screechy, nails on a chalkboard style) band, and/or various interlocutors*: mystic Tomato, mystic tomato (*typically with more harmony and/or taste than herb's first four words*)

> *herb then usually follows with a diatribe about the tax man, or as most prefer, a rendition of the beginning of the shaman Neil Young's 'After the Gold Rush'*

herb: i Dreamed I saw the knights in armor coming saying something about a Queen; there was a peasant singing (*herb himself*), Bobbie strumming, and an archer split the tree; there was a fanfare blowing to the sun, and I felt like getting high... Look at Mother Nature on the run in the 21st century, look at humanity under a boot in the 21st century; and eye say damn Zeus's tyranny, damn Zeus's tyranny, damn Zeus's tyranny on the affairs of wo/man.

> *after this short introduction, herb always announces the band in the same way...*

herb: up Next, Theurgist of Atlantis, Live from the Sawmill Saloon, B// R///, planet Earth, Milky Way Galaxy, physical Universe, the lowest realm in all of god's creation

ACT II

on Physics flow Of that mysterious fluid type magical substance called electricity flow Of number thats So meta chips And dip on the waiter deck a Solution in time a Grand tour a Reality flux fate Fate fate universe As computation computer As scratch pad the Obvious analogy the Light Life in of plants ours Versus it ode To an ancient idea variations Of the computer theme behold A simulation aether computational Pleroma on Computational steering hail FIFA power Source everywhere and everywhen et Tu (physical) universe seething Dance in a cave of shadows the Data of universe carriers Of force the Code of universe some Subtleties an Elephant in the physicists room on The nature of gravity and Now what a Dance before The physical little K-nots of 'light' on How nature computes the Light is in the playground principle Of no local holding patterns

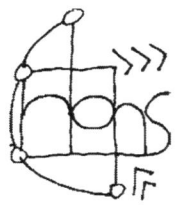

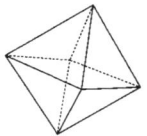

INNER-MISSION

on Mind the Fable of the simulation argument a Functionalist masquerade eyes Wide shut u R living in a computer simulation what Is a posthuman to compute? mechanisms Four the simulation argument observation Selection effects and the nature of time nomological features and the enslavement by time a Fuller gravity gnostic Theology the Brain builders dream on the flexibility of time and the need for the divine

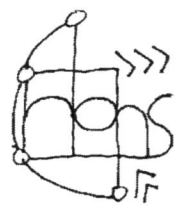

on Physics

Do what thou wilt shall be the whole of the law.
- Fra. P.

> *the interlocutors, herb, Giordano, (the goddess) Sophia, Joe, and AlBe are walking in a midday sun, when Giordano poses a question (and a damn good one to be sure)*

Giordano: my dear herb, is it possible, in your humble opinion, to contemplate an ontological reality associated with Universe?

herb: thank you my dear Giordano, for such a lovely and probing question. Quite fitting for such a fine day as today.

To start then, we should acknowledge what is meant by the word 'ontology.' To wit: ontology is the metaphysical study of the nature of being and existence. Quite delightful this study of ontology then must be, if in fact it purports to be *the* field of study about all that is (whatever is is).

Sophia: i am so glad you qualified this at the end my dear herb. But please, before you proceed, please advise us on the proper interpretation of the word metaphysical, which you just used to start to explain ontology.

herb: with pleasure, my dear (the goddess) Sophia, as the words we use are apparently the 'things' that bring forth 'thoughts' to our awareness.

Giordano: what does that mean, are you saying that words are a type of magic?

herb: consider then my dear Giordano, it is quite a 'physical' system in play that allows us humans (and you k-now who you are) to understand words, and if we were to spend a complete afternoon contemplating it, the actual mechanisms involved in that flow of words which begets a flow of thoughts, I mean to simply say, the mechanisms that support our human (and you k-now who you are) bodies ability to 'hear' a sequence of sounds, and then summon past experiences and knowledge to - evidently - receive some type of picture show that actually has something in common with the picture show(s) of others - all this simply because we heard the same 'word,' is very extraordinary would not we all agree?

Sophia: and why talk about this, when we are on the topics of ontology and metaphysical?

herb: while I want to make sure we get a good understanding of ontology, meaning, I suppose, that we should get a type of consensus on what particular 'thoughts' should be summoned when we use the word 'ontology,' I will proceed as you (the goddess Sophia) suggested, and begin by considering the word 'metaphysical.'

Sophia: thank you my dear herb, and can I add, that it is agreed by most that words, especially these with agreed upon meanings, do act as a sort of magical enhancement, in that the thoughts brought forth - within / wherein? - to ones' mind is just so fantastically consistent (once one has learned the language that the words are spoken in).

Giordano: i thought (hehe) we were going to talk about what meta-physical meant!

herb: thank you my dear Giordano for putting us back on track, and thank you my dear (the goddess) Sophia for always being willing to help me elucidate the issues and meanings involved in our philosophical wanderings.

flow Of that mysterious fluid type magical substance called electricity

> *Joe offers herb a smoke, and suggests he give an example selected from things in this physical Universe, to re-introduce what he is trying to say about ontology and metaphysical*

herb: consider the ancient idea of a computer system, if you will, where we have programs running on the computer. The programs purpose in the final analysis is to control the flow of that mysterious fluid type magical substance called electricity. The complete elaborate system is setup so that the program code that we create first gets translated into a series of bits[37], and this series of bits associated with the program code is responsible for enticing the emanation of control signals in the circuitry of the computer. These control signals control the flow of that mysterious fluid type magical substance called electricity in the circuitry of the computer in such a way to create a meaningful final result of the computation. For example, in a program designed to calculate the square root of the number 2, the program code represented in the memory of the computer controls the actions of the computer (and the associated flow of that mysterious fluid type magical substance called electricity) in such a way that the final result of 1.4142136 is obtained (and displayed).[38].

So then, even though it is / are physical events that transpire inside the computer, we have abstracted away from the flow of that mysterious fluid type magical substance called electricity, and as programmers we can think instead of the logical operations and a time sequence / ordering of operations involved, in, say, the calculation of the $\sqrt{2}$. We can do this kind of abstraction with physics also.

Now then, if we take the word physical, this word can bring forth thoughts of physics (that is, the mathematics that help us understand and quantify the natural physical world). And if we agree

---

[37] a bit is the name for a single digit of a binary number. A binary number is a base 2 number, as compared to a decimal number, which is a base 10 number; thus a decimal system has ten digits, 0 - 9, while a binary system requires only two digits, 0 and 1.

[38] it is funny here that herb chose the $\sqrt{2}$, which he was often fond of reminding people was not a real thing at all, since the digits after the decimal point never resolve, so no 'thing' could ever have a length of $\sqrt{2}$. Further, from the Pythagorean Theorem, $a^2 + b^2 = c^2$, one can surmise that there are no physical flat surfaces, since for a square box with sides length 1, one would have a diagonal of length $\sqrt{2}$, but there is no such thing as the $\sqrt{2}$, ergo, there can not be any flat surfaces in Universe... and this use of the $\sqrt{2}$ that herb often recited was patterned after a similar, if not, exact way in which the great shaman R. Buckminster Fuller explained it

that there is such a thing called physics (and this includes all the mathematical and geometrical reasoning that can be employed, and all the physical experimental equipment that is built and / or conceived that allows us to make measurements to test said mathematical and geometrical reasonings), then we are left with the prefix *meta*, in order to explain the term metaphysical.

flow Of number

Joe and Giordano look at each other, both likely hoping that herb will eventually get to the point

herb: in order to get the proper picture that allows us to undertand deeply the term metaphysical, I want us to stay with this computer system analogy, wherein we have this flow of that mysterious fluid type magical substance called electricity. We then replace one flow (of that mysterious fluid type magical substance called electricity) with a viewpoint / abstraction that it is a flow of number that is taking place (when a program is running on a computer).

Sophia: my dear herb, so you now have us viewing what is a physical process, i.e., the flow of that mysterious fluid type magical substance called electricity (in the circuitry of the computer), on at least two more levels, and this recent level I must say is the most beautiful. You have helped me create a vision that *is* a flow of number, but at the same time seems to be an elaborate dance (at incomprehensible speeds) of that mysterious fluid type magical substance called electricity. However, if given a choice between these two views / abstractions, it truly is number that flows (whatever number is?)[39].

Giordano: does the programmer who writes the program which runs on a computer ever have to think in terms of the two different flows that you and (the goddess) Sophia are speaking about?

---

[39] The great shaman Pythagoras often said that all things are numbers.

herb: no my dear Giordano, and rarely, I would guess, does a typical programmer consider the actual flow of that mysterious fluid type magical substance called electricity when they are waiting for their program to finish running. Irrespective of the programmers thought process however, the events that transpire within the circuitry of the computer must be completed in order for any result to be computed.

Joe: pardon me my dear herb, what does this (beautiful) idea / abstraction of the flow of number have to do with the word metaphysical?

thats So meta

> *the interlocutors are joined by thalamus and Glia, who are simply delighted to have happened upon a gathering where thinking is allowed. herb attempts to keep the party going (cough cough)*

herb: please join us as we continue to attempt to elucidate the meanings of the terms ontology and metaphysical. I have set the stage for an explanation of metaphysics by whimsically (and covertly) trying to develop a physics about the physics inside a computer during the running of a program. We have, I believe, agreed that we can view the running of a program in at least three ways: the flow of that mysterious fluid type magical substance called electricity, or as a flow of number, or as a sequence of logical operations and program statements created by the programmer.

> *herb instructs thalamus to ask Giordano to explain flow of number to him at our next break. Joe looks at herb and says the word: metaphysics*

herb: consider this: We have a language, which, say, is used to communicate thoughts / feelings / divinations of the future /

reflections of the past / and all other wyrd things that are involved in our speaking and writing. Then, what if we wanted to create a language that would be used to analyze language; that is, we are trying to create a *language about language*.

Sophia: what should we call this language about language?

herb: we would properly call our new language about language a metalanguage. Question: what then would we call another level of language that we could build to analyze / contemplate / think about metalanguages?

Glia: a metametalanguage?

herb: correct.

Giordano: so following this line of reasoning, we would say that metaphysics is a physics about physics?

herb: to properly think about physics, we could indeed be assisted if we had a physics about physics, which as you have pointed out my dear Giordano, is aptly called a metaphysics.

chips And dip on the waiter deck

> *Joe has been pulling along with him a cart (more of a wagon) with some supplies and party favors, as he k-nows well that a day of thinking makes one rightly hungry and thirsty. thalamus and Giordano will soon slip away to talk about the flow of number*

herb: our dear Giordano posed a somewhat daunting, but delicious question concerning our ability / or lack thereof to contemplate an ontological reality associated with Universe. It was during a definition of the word ontology that we became concerned, and rightly so, with a definition / understanding of the sound sequence that we write down as metaphysical. In turn, the discussion of metaphysical lead into a needed understanding of the use of *meta* as a prefix. We have successfully done that, but now I feel we must talk about the physical some more, because Giordano's question not only contained the word ontology, but it was chiefly concerned with whether or not the physical could be ascribed an ontological viewpoint.

Glia: so how should we continue then?

herb: if it is agreeable to all, I would truly like to take us on an adventure in thinking which I would now suggest will elucidate many many fine and subtle things about our physical world.

AlBe: are we to get a lecture on physics? one of your favorite subjects!

Joe: should we have a meal before we get started on this aforementioned adventure my dear herb?

herb: i am quite thirsty, any adult beverages in that wagon?

a Solution in time

> *it was after the meal, and potty breaks, that the interlocutors were ready to proceed again. thalamus and Giordano have returned, and they are laughing and grinning, and exude an aura that the divine may envy*

herb: i propose to use the idea of a time-domain simulation, run on a computer, to both motivate and inform our discussion on physics. We will take that nature herself is a sort of computation. In doing so, I hope to build a deep analogy between the way of our ancient computing techniques, and the way of the natural world.

Sophia: i am so sorry to interrupt my dear herb, but can you tell us what is a time-domain simulation?

herb: a time-domain simulation is a program that simulates some physical event(s), by advancing a 'solution' in incremental time steps. The 'solution' is in a way a complete description of the physical events being simulated. We say 'time-domain' because each version of the 'solution' represents a certain point in time.

Giordano: sounds like an interesting adventure my dear herb, can you give a quick example?

herb: if I drop a coin (say, one with King Og's face on one side of it, and his arse on the other side of it) from a height of 6, we do currently k-now the basic rules of how to calculate both the velocity and position of the coin. At time 0, the coin is at a height of 6, and sometime later the coin will be at a height 0 (when it lands on the ground). We can then, if we so desire, step the 'solution' (here the solution represents everything about the coin: height, velocity, etc.) in increments of milliseconds or microseconds.

The complete time-domain simulation of the coin dropping would produce a 'solution' that would track the coin from the height of 6, and at (usually) pre-chosen increments, give another version of the solution at some time α, where it could have a height of, say 5.5 . Then there is a version of the solution at time 2α ($=\alpha+\alpha$) where the coin is at a height of 4, and this continues on until the coin hits the ground (which happens for this example between 3 and 4 α's).

thalamus: what if the coin was rotating in the air? as in a coin flip?

herb: yes my dear thalamus, we could add descriptions and mathematics (and computer code) for all the dynamics of the coin in motion, as we do have very good techniques for analyzing rotating objects (using Euler angles, for example).

Sophia: so the level of detail one puts into the so-called time-domain simulation is up to the programmer; based partly one would suppose, on what kind of techniques are in their toolbox.

a Grand tour

> *Jove has propelled (the goddess) Fate forth to locate and join the interlocutors. Giordano and (the goddess) Sophia will most likely delight once again in her divine counsel*

herb: so it is agreed, we will use time-domain simulation of physical events in Universe to both motivate and inform a discussion on how nature computes.

Giordano: and then a byproduct (or side-effect) of this discussion will be a deeper appreciation and understanding of an ontological underpinning of 'things'?

herb: precisely my dear Giordano, as we will build a deep analogy between parts of the simulation and the actual 'things' that are being simulated.

Sophia: but the operation of the computer (during a time-domain simulation) seems nothing like what nature is doing at any moment in 'time.'

herb: quite the contrary my dear (the goddess) Sophia,

> *this surprises herb, that (the goddess) Sophia would make such an oversight... unless she (the goddess) was playing a bit of a game with him*

and important attention need be paid to the idea of what I will label as *computational steering*, and the fact that computation requires *power sources available throughout* the circuitry (of the computer) during a time-domain simulation.

Giordano: herb, (the goddess) Sophia, a beautiful sight appears on a far hilltop. If I am not dreaming, I will tell all who would hear, that (the goddess) Fate has appeared, and in a short time, I suspect, will join us here.

a Reality flux

> *herb continues the Grand tour, and will soon welcome (the goddess) Fate*

herb: so then, with these ideas of computational steering, and power sources everywhere, we will - naturally - move to consider Universe as a computation (of sorts), with these same type of requirements, or parts (if you will).

thalamus: does the Universe also use bits for it's computation my dear herb?

herb: my dear thalamus, many do look at the goings ons in Universe as strictly informational, and quite often do render this type of thinking in terms of bits.

thalamus: and this is not how we will approach it?

herb: no my dear thalamus, we need go down to what some have termed the Planck scale[40], and contemplate the idea of a *quantum foam* - an idea relayed to us by the shaman Wheeler.

Sophia: what do you mean my dear herb when you use the word *quantum*?

herb: quantum is a term that involves most likely two basic ideas. One, from the great shaman Planck we k-now that energy is quantized into *packets* at the lowest levels. For example, a 'particle' of light that we call a photon, at a frequency, f, has an energy value of hf, where h is (the famous) Planck's Constant.

The second important idea associated with quantum is the so-called *uncertainty principle*, which can be stated mathematically as

$$\Delta E \Delta t \geq \hbar/2, \text{ or } \Delta p \Delta x \geq \hbar/2,$$

where $\hbar$ is Planck's constant h divided by 2π, and E is for energy, t is for time, p is for momentum, and x is for location (or displacement). And, while the uncertainty principle in one sense places restrictions on how accurate measurements can be, it also allows (nay, requires?) the involvement of the so-called *quantum vacuum* field[41] in physical activities in Universe.

Sophia: so if that is the basics of quantum, what then do you mean when you say quantum foam?

herb: my dear (the goddess) Sophia, let's consider what humans (and you k-now who you are) see in the natural world: other humans (and you k-now who you are), animals, birds, mountains, rivers, et. al.. Then, we hear that all these 'things' are composed of molecules and assorted atomics, which are themselves composed of elementary particles and so on.

---

[40] T The Planck length, and the Planck time are very small quantities that are supposed to lie at the bottom of the goings on's that is the Universe as computation.
[41] also called the zero point field (ZPF), or simply the quantum vacuum.

The mystery of scales in the realms of these players are of course fantastic indeed, which one could get a glimpse of with the help of different types of microscopes and the physical theories. Combine these realms involving humans (and you k-now who you are) and below with the (physical) realm above involving the solar system, the galaxy, and other galaxies into the large Universe of stars, and the wonder of scale is breathtaking.

Joe: quantum foam herb?

herb: right then; at the bottom level of our physics understanding, say, below the particles which make up the atomics, is the Planck scale realm, and it is here that the quantum vacuum mechanisms run the rule. For example, we find in the physical theory that physical processes need 'things" to jump in and out of existence. It is also conjectured that space-time itself is being manipulated down here at the lowest level, a playground that the shaman Wheeler aptly named the quantum foam.

Sophia: so then the quantum foam is where 'things' jump into and out of existence? Where do 'things' 'go' when 'they' are not to exist anymore?

herb: my dear (the goddess) Sophia, I will provide a name for the abode that you refer to in due course, but let us all focus now on the requirement itself.

That is, a major part of the quantum vacuum mechanism is that we have 'things' popping into and out of existence, and these 'things' are in fact required for any process at all to occur in physical Universe; and I call this activity the reality flux. Further, we will conjecture that this *reality flux* will act as the power source for a Universe as computation.

fate Fate fate

> *herb welcomes (the goddess) Fate to this days gathering of interlocutors. As the sun will set soon, the interlocutors are off to an old part of the river, where AlBe has some snacks readied*

herb: my dear Giordano, as we all together move to the ancient park by the river, would you be so kind as to catch (the goddess) Fate up on the conversation of the day.

Giordano: my dear (the goddess) Fate, I welcome you to our humble gathering, and it is my task now, and my pleasure, to bring you to an understanding of why herb is talking about ancient computing techniques.

Fate: word is that you, my dear Giordano, desired to k-now if one could identify an ontological reality that would describe and help us understand Universe... up to your old tricks I see.

Giordano: you k-now me well (the goddess) Fate, and you are correct in how you describe our current quest. herb has given much background information in hopes that we would understand why he is to build an analogy between the ancient way of the computer and the way(s) of the natural world.

herb: and, we take notice if we dare, that the k-nots that 'happen' with that mysterious fluid type magical substance called electricity are indeed the best place to put our ontological reality that we seek, and further, these k-nots along the strings of that mysterious fluid type magical substance called electricity are like the threads managed and manipulated by the three fates of fable; if it pleases (the goddess) Fate that I propose such.

Fate: it is indeed pleasing my dear herb. Now what about those ancient computing techniques, and the ways / fates of the natural world?

AlBe: are those 'fates' of fable the same as those three (3) witches in that ancient control manual called MacBeth?

Giordano: indeed my dear AlBe, the 3 fates that MacBeth conversed with are those same of fable that herb alludes to.

Sophia: right.

universe As computation

the interlocutors have arrived at a very old park by a bend in the river. If one allows oneself to listen closely, an ancient song or three can be heard swirling in the air.

herb: right then; after we have covered all the basics, I would hope to revisit the possibility of an analogy between the 'things" / stuff of the physical Universe and various parts of the computer that exist during the running of a time-domain simulation. At that point we should be ready for an overview of a geometrical computing style called computational Cosmography, and we can at that time see how this ideal matches up (or not) with the way(s) of the physical world.

Sophia: so if Universe is a computation (of sorts), then your claim is that the reality flux part of the quantum vacuum acts as the power source for this Universe as computation. But what of the other 'things'' / stuff in (physical?) Universe that do not so easily give in to algorithmic reasoning?

herb: thank you my dear (the goddess) Sophia, as these extra 'things' '/ stuff that do not so easily give in to algorithmic reasoning should also be included in our discussion. For example, we will indeed ponder how *Mind* can operate in such a computation, and for this to occur (i.e., Mind in Universe), we will see the need for something else besides just the physical.

Sophia: i suppose at the appropriate moment you will then bring into play many of those Gnostic nasties that so enrage so many people, but at the same time make so much sense when one is privy to the real history that is this physical Universe.

Fate: but, my dear (the goddess) Sophia, it would only be appropriate to consider the ancient Gnostic wisdom when one starts to contemplate how Mind can intervene on the physical.

herb: and that will be a somewhat related adventure, surely saved for another time, after, say, an inner-mission.

computer As scratch pad

the next day the fire pit is still going from the festivities of the night before, and the interlocutors take turns with the utensils that are used to prepare their morning meals. A supply of juices and coffee beans is not yet exhausted, and laughter is heard up and down the river after a good joke has been told

Joe: my dear herb, for those of us not so familiar with the ancient computing techniques, can you explain what does go on inside the computer while it is running a program, or specifically, a time-domain simulation?

herb: the computer in the most basic sense acts as a type of scratch pad; a place to keep track of things. Consider the following diagram

herb draws on a stone table - using chalk - a diagram of a basic computer setup, including the memory, the CPU, and a display

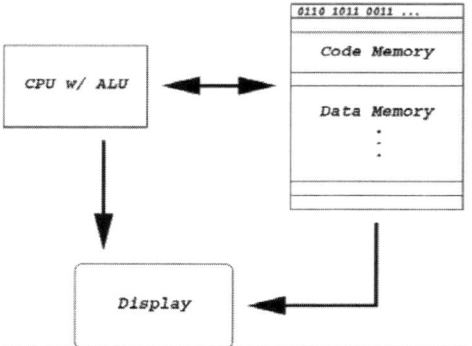

Here in memory, the bits associated with the program code are intended to instruct the CPU (i.e., Central Processing Unit) how to manipulate the bits in the data memory.

Sophia: are these bit manipulations fairly random my dear herb?

herb: no, my dear (the goddess) Sophia, for example, in a time-domain simulation of something of interest in the physical world, the bits that make up the program code organize in a very particular way how the bits in the data memory are to be manipulated.

The manipulation is pulled along in such a way so that the data memory resembles (after proper rendering of the information, say, on a display) what is going on with the 'thing' being simulated.

Joe: it all sounds quite elaborate indeed.

the Obvious analogy

(the goddess) Fate is smiling as Giordano and thalamus also light up

herb: now, the obvious analogy that I need point out here is that the *data* (memory) of the time-domain simulation resembles the physical 'things' in Universe, while the *code* (memory) of the time-domain simulation acts in ways similar to how the physical laws (in Universe) act.

Fate: so it is the code that instructs the data what to do, just as it is the physical laws (in Universe) that instruct the 'things' in Universe what to do.

Sophia: my dear herb, is this not also how we think of electromagnetic energy events (associated with that mysterious fluid type magical substance called electricity), in that it is the Maxwell Equations (here, the equations would be the *code*) that dictate what the electromagnetic energy (the *data*) is to do in each successive instant of time?

herb: yes, this is exactly correct. In fact, in a computational electromagnetic time-domain simulation, this analogy is highlighted even more. In this type of simulation we setup data structures for the values associated with the electric field and magnetic field values.[42] The *code* of the time-domain computational electromagnetic simulation will act in an analogous way to what happens in the physical world, in that the code will provide the effect of the so-called Maxwell equations[43], and coordinate the manipulation of the data, so that the data resembles - properly - the correct electric and magnetic field values for the given situation being simulated.

thalamus: can any type of electromagnetic situation be simulated? For example, could one simulate the electromagnetic interaction between plants and 'light'?

---

[42] The electric and magnetic field values are a convenient way to abstractly view, and quantify, the mysterious fluid-type magical substance called electricity.
[43] our current best description of how the mysterious fluid-type magical substance called electricity evolves in time.

the ~~Light~~ Life ~~in~~ of plants

> *(the goddess) Sophia and (the goddess) Fate briefly discuss whether some things should always remain unspoken, or whether or not it is permissible to ponder anything whatsoever*

herb: my dear thalamus, this is such a keen observation you have made. In fact, the interaction between plants and 'light'[44] is that famous process of photosynthesis, wherein plants literally 'eat' light.

This would be a grand time-domain simulation indeed, but let us step back (and maybe aside) to consider this photosynthesis process in its own right, as it may enlighten us as to another aspect of that mysterious fluid type magical substance called electricity.

Sophia: just as the circuitry of a computer is a set of pathways for the flow of that mysterious fluid-type magical substance called electricity, does not a typical plant also act as a conglomeration of circuitry for the flow of that mysterious fluid type magical substance called electricity?

herb: indeed my dear (the goddess) Sophia. The symbiotic relationship between light and plants is like that, and much much more. For me, I like to think about this interaction between plants and light in terms of the flow of number / light that flows in the plant, and which in turn produces many other wonders.

Fate: it is within this consumption of that mysterious fluid type magical substance called electricity that the creation of (beautiful) molecules within the plants are also accomplished. In a way, the flow of number begets new types of numbers, and the flow of number (that is, the 'light') also, in this case, also seems to power the machinery of the plant(s).

herb: it would seem, that when one considers things in their simplest sense, that it is this flow of that mysterious fluid type

---

[44] 'light' is a form of electromagnetic radiation, which in turn is just another name / term for a version of that mysterious fluid type magical substance called electricity.

magical substance called electricity inside plants that forms the foundation of this planets ability to sustain life forms[45].

And before I (try to) move back to talking about the actions / activities inside a computer during the running of a time-domain simulation, I would like us all to consider the differences between the mysterious fluid type magical substance called electricity that we use in modern machinery (and in the circuitry of a computer), and the instances of that mysterious fluid type magical substance called electricity that are found (naturally?) in the physical world (and, which, as recently discussed, power the machinery of plants, for just one example).

ours Versus it

> *the interlocutors all take a coffee break, while a few also go for a smoke. Visions of the flow of number inside plants occupy many of the minds, some pondering what a fine web is woven by this light found in the natural world, and some even dare consider that that is all there is (whatever is is)*

Giordano: my dear herb, is there a distinct difference between the forms of that mysterious fluid type magical substance called electricity that are used to power the machinery of us humans (and you k-now who you are), and the different types of that mysterious fluid type magical substance called electricity that we find all around us in the physical world?

herb: my dear Giordano, while all versions / instances of that mysterious fluid type magical substance called electricity can be described fairly well by the famous Maxwell equations[46] I think at this time it is important to describe how we humans (and you k-now who you are) harness for our own use various versions /

---

[45] which may provide testimony to the rationale behind the ancient sun worship cults.
[46] here herb means to say that various measurements of most situations involving that mysterious fluid type magical substance called electricity match up well with the predictions that result from using the Maxwell equations to predict the time-evolution of the electric and magnetic fields associated with any particular electromagnetic event.

instances of that mysterious fluid type magical substance called electricity.

Sophia: is it not through the use of spinning magnets that we harness / acquire a time-varying version of that mysterious fluid type magical substance called electricity my dear herb?

herb: indeed, my dear (the goddess) Sophia. It is through the ingenious techniques that involve rotating equipment that we humans (and you k-now who you are) basically (in no uncertain terms) *summon* that mysterious fluid type magical substance called electricity directly from the 'aether' of space.

thalamus: is this what is happening when we use water flowing out of the bottom of a dam, or in a large river, to create 'electricity'?

herb: yes my dear thalamus, the water, under great pressures, turns turbines around and around as the water flows by, and it is during this rotation of magnets, which are surrounded by wires, that what you term 'electricity' suddenly appears in the wires wrapped around the magnets.

thalamus: in electric power plants that are not built with dams or over rivers, how is 'electricity' generated there?

Joe: if I may my dear herb; for these situations, we still use spinning magnets, but a different energy source is used to (typically) boil water, and then it is the steam produced from the boiling water that turns the turbines, wherefrom, again, the 'electricity' appears in the wires surrounding the magnets - and is then distributed forth over a series of wires for consumption by various machines.

Fate: this sounds all so magical in its operation my dear herb, in that, it seems that that mysterious fluid type magical substance called electricity must be in a sense hiding and waiting for the turning of the magnets in order to escape from some other realm and then enter this physical realm whenever some of the magnet-based equipment starts to rotate.

herb: indeed my dear (the goddess) Fate, and it is believed that wherever or whenever such magnet-based equipment is turned, that mysterious fluid type magical substance called electricity will appear.

Sophia: and you mentioned in passing earlier my dear herb that this spontaneous appearance of that mysterious fluid type magical substance called electricity in these wires of the oft-described rotating equipment is a type of summoning from the aether of space... please tell, what is this *aether* of space you refer to? Is it an abode of a type for that mysterious fluid type magical substance called electricity... and also an abode for things of the quantum foam one would have to presume

herb: if presuming is allowed.

Glia: it most certainly is.

ode To an ancient idea

> *(the goddess) Fate casts a wyrd gaze towards (the goddess) Sophia, knowing since the beginning of time her involvement in these matters, all the while wondering how Gnostic / jinkies herb is going to get with it - the aether - a fabled substrate - the Pleroma in the ancient mysteries*

herb: it was said concerning the great shaman Maxwell, that it was impossible for him to think incorrectly about physical phenomena. The great shaman Maxwell envisioned some type of substrate (if you will) existing throughout space, and that that mysterious fluid type magical substance called electricity, in the form of traveling electromagnetic fields, was supported in its propagation by the / some actions of this substrate.

Many at the time called this substrate (of the entire physical world) the aether. In later years after Maxwell, it became, for some quite unreasonable reasons most likely, quite unfashionable to believe that there was such a thing as an aether.

Sophia: but is not the quantum vacuum, a presumption of most scientists in recent time, a type of aether? Is it not assumed to be everywhere (and everywhen), and is it not presumed that the quantum vacuum supports the interaction of all 'things' in Universe?

Fate: and what of that famous hoax concerning the so-called Higg's Field, that fabled field that permeates all space, and which in its actions facilitates all matter acquiring mass?

herb: you are both correct, in that both the quantum vacuum and the Higg's Field should be counted properly as a type (or aspect) of an aether; for they play the same role as the ancient aether ideas (which sometimes went by other names).

Giordano: my dear herb, how does this idea of an aether compare to the ancient idea(s) of a Pleroma, a quasi-material type thing-y within / wherein all 'things' are, in a way, floating (for a lack of a better term at the moment)?

herb: maybe to say 'embedded' would be better than 'floating'?

Giordano: embedded indeed.

herb: my dear Giordano, I would submit that the great shaman Manly P. Hall, in his seminal "Secret Teaching of All Ages," would rapidly agree with you and I when I say the ancient Pleroma is now called by science the quantum vacuum (even if not all the attributes of the ancient Pleroma are properly associated with the modern quantum vacuum - yet).

variations Of the computer theme

> *it was after the interlocutors queried both (the goddess) Sophia and (the goddess) Fate about the ancient Pleroma idea that the interlocutors decided to go fishing, in hopes of catching an evening meal*

herb: so I have tried to get us to view the data and code in the memory of the computer in an analogous way to the 'things' of the physical world, and the associated physical laws, which in a way guide said 'things' along their time-domain paths according to our best 'physics' interpretations. Now I would like to consider some modern variations of the basic 'computer as scratch pad' idea.

Joe: so remind me quickly then herb, you called the memory of the computer a 'scratch pad' because the data memory was continually being modified via actions dictated by the code memory - is that right?

herb: right Joe, and at this time I wish to point out that each of the different sections of the data memory (analogous to 'things') needs to take turns being updated, as only so many things can be manipulated by a CPU at any one time.

thalamus: but was it not the case that many of the computers had many CPUs working together?

herb: indeed my dear thalamus, and this is a type of variation on the basic theme. For a so-called shared-memory multiprocessor had many a CPU all working on a common shared scratch pad (updating the data memory). Here then the data memory could get more attention that usual when there was only one CPU.

Another variation on the basic theme were distributed-memory multiprocessor systems, which would actually partition the data memory into separate partially disjoint scratch pads.

Fate: how do these variations of the basic 'computer As scratch pad' model compare to how we view 'things' in physical Universe, and the associated laws which govern their behavior?

herb: my dear (the goddess) Fate, it is implicitly assumed that all 'things' in Universe get constant attention from the physical laws; that is to say, 'things' in Universe most likely do not take turns getting their instructions for how they are to continue on along their course of destiny in association with necessity. These basic computer variations do allow the individual pieces of the data memory to get more attention than before, but it is not a continuous attention that data receives in these models.

behold A simulation aether

> *many a fish and berries are being prepared for a lovely meal. After the meal, and before the evenings music, the discussion continues*

herb: also please note that the data memory within the 'computer As scratch pad', when running a time-domain simulation, is written back into a backup storage area almost constantly during the running of a typical program.

thalamus: and is this the case because the complete simulation can not fit into the data memory area?

herb: that is sometimes the case, and it could also be due to some other choices taken by the operating system that 'officially' runs the computer.

Sophia: so then the physical 'things' that are being simulated, sometimes are put into computational hibernation?

herb: yes, when it is not the 'things' turn to be computed about.

thalamus: are their other computational scenarios where 'data' can get more attention? say, constant attention?

Fate: as it is with all 'things' in (physical) Universe.

computational Pleroma

> *herb wants to describe a computation style that resembles a simulation aether, or a computational Pleroma*

herb: yes, as (the goddess) Fate k-nows well, 'things' in (physical) Universe seem to 'always' be subject to physical law.

Giordano: all the way up and down the mystery of scale; that is, down to the quantum foam, and up to the being of galaxies and planets (including our Jove).

herb: and so the idea came about that the 'data' should be placed directly within the logical circuits which perform the necessary updates of the data - and in this way the circuits of the device have the effect of 'code' memory and the running of the code on the computer to update the data.

thalamus: and since the data is now in a way embedded in the circuits, the data gets constant attention?

herb: attention with every 'clock' pulse sent to the computational Pleroma (circuit).

Fate: and so the 'data' gets constant attention from the circuits, and thus the circuits in this setup act as the physical laws (that tell 'things' what to do).

Joe: right.

on Computational steering

after the meal of fish and berries and beverages, a band plays into the night before each interlocker drifts into their individual dream worlds to ponder and dance. The next day while enjoying coffee and smokes, some of the interlocutors ramble on

herb: another key point about a time-domain simulation being run on a computer is how to 'steer' the simulation correctly based on the values the 'data' take on. I will try here, now, to make the point that any 'computational steering' of the simulation (running on a computer) comes from 'outside-the-system' proper.

That is, the computational steering of a simulation is typically accomplished in one of two ways: either (1) by actions / controls placed in the 'code' memory, or (2) steered interactively using 'from-outside-the-system' inputs.

thalamus: but my dear herb, you first said that 'all' computational steering of a simulation comes from outside the system, but only your second example here seems to use 'from-outside-the-system' inputs.

herb: indeed my dear thalamus, at first blush it appears that way, but when the manipulation of the data memory is solely done by elements of the program in code memory, I will claim this is still a result of 'from-outside-the-system' inputs.

Glia: my dear herb, are you claiming this because the code memory itself is a result of a program written (typically) by a human (and you k-now who you are), who - sans a Tron-like character - is outside-the-system?

herb: indeed my dear Glia, this is exactly my point here. The decisions as it concerns the time-evolution of the data memory are taken by the code memory, which is a result of a program written by an entity from outside the system.

hail FIFA

> *more interlocutors have joined the discussion, and some are excited that the talk seems to have turned to football*

Sophia: my dear herb, what about those famous football games, played with a controller, where one can manipulate what look like real players on a football pitch? Is not this a good example of where inputs to a time-domain simulation come 'from-outside-the-system' inputs?

herb: game systems of old, my dear (the goddess) Sophia, should be viewed as time-domain simulations proper to be sure, if the dynamics of the games include physically based motion.

thalamus: like the FIFA12 game where I press an X for Ronaldo to shoot, press an O for him to pass (rarely), and a Square to cross!

herb: indeed my dear thalamus, exactly such. This is a great example of a steered time-domain simulation that has data to represent the current state, and algorithms turned into code that control the setting of the next state of data, based on the current data and any of the possible 'from-outside-the-system' inputs (coming from the controller(s)).

power Source everywhere and everywhen

> *the interlocutors have moved on from their camp by the river and the next midday sun brings them upon some ancient, seemingly not k-nown before caves. (the goddess) Fate lights the way*

herb: we have talked at length about the ways of ancient computing techniques, wherein, can I remind all, the flow of that mysterious fluid type magical substance called electricity is managed in such a way that a result can be calculated / produced. Butt now, I need point to an oft overlooked fact about computers, which involves the circuitry of the computer, where in fact power sources and ground connections need be available throughout the circuitry.

Joe: right, all those logical and sequential circuit operations performed using transistor-type circuit arrangements always required access to a HIGH voltage source (to produce a binary 1) and access to a ground connection (to produce a binary 0).

Sophia: and so even though we can think of computation as a flow of number, it is not like a flow of a river, where, say, the amount of water across any cross-section at any one point is more or less the similar to the amount through another cross-section two clicks down river.

herb: my dear (the goddess) Sophia, indeed, it is not like a river, for if a river were like computer circuitry, we would require incoming flows every step or so along the river and would also have almost as many sinks (where the water flows out of the river) every step or so also.

et Tu (physical) universe

> *(the goddess) Sophia decides to change her tune (to more of a classical one, using a divine flute, if one is able to imagine such a thing), and as the interlocutors come to the end of a Grand tour (and to an entrance to a cave), (the goddess) Sophia agrees with herb (publicly) that there may be something to this analogy*

Sophia: my dear herb, this requirement you point to for the computer circuitry, that is, the need for a power source everywhere, is not this very similar to what you have pointed out about the quantum vacuum mechanisms, that in fact those mechanisms are required everywhere and everywhen for any process at all to occur in physical Universe?

herb: very well said and quite insightful, and thus welcome aboard (the goddess) Sophia (and nice flute by the way). This is indeed part of the analogy we are altogether trying to build and understand. The analogy between the way of old computing techniques and the way of Universe (as computation). For it is as you have supposed, the reality flux part of the quantum vacuum mechanisms acts as the 'power source' for Universe as computation in an exact manner that voltage / ground / current act to power the circuitry of computers.

Giordano: and the rest of the analogy, so far, if I may, my dear herb, is that the data of the time-domain simulation is analogous to the 'things' in physical Universe, while the code of the time-domain simulation is analogous to the physical 'laws' us humans (and you k-now who you are) have uncovered about how Universe manipulates 'things' ... as time rambles on.

Joe, herb: right

seething Dance in a cave of shadows

> *with the option of many paths, as is usual in the life of a seeking fool - the divine willing, is also the situation the interlocutors find themselves in as they emerge into this particular labyrinth. (the goddess) Fate continues, it seems, to light the way*

herb: before moving to describe the 'physics' of Universe as seen by 'modern' science within this analogy we are building, we should point out that it seems to be that Universe only maintains one copy of each 'thing' (or energy event) in physical Universe as the physical Universe evolves in time.

thalamus: whereas in many computing techniques, there may be different versions of the 'data' spread about the memories of the computer.

herb: butt it could be my dear thalamus that Universe has 'another version' of 'things' also, in what I like to term

'Negative Universe'[47] (which can be envisioned as a type of non-spatio-temporal[48] abode). For a lack of a better way of saying it, we can think of Negative Universe as the abode on the 'other side' of the reality flux.

thalamus: so that the things of the reality flux, which as aforementioned, acts as a power source for Universe (as computation), are interacting between a version of a 'thing' in physical Universe and a(n) (ethereal) version of said 'thing' 'in'[49] Negative Universe?

herb: and before we continue to investigate this possible relationship, lets step below to understand better what is happening in physical Universe proper.

Sophia: are you now, my dear herb, to take us down to the quantum foam that you and others have speculated runs below all of physical existence?

herb: and right you are (the goddess) Sophia; for when one really wants to contemplate Universe (as computation), we should in one sense view physical Universe as a seething dance of energy events whose true dynamics should be couched in terms of activities at the so-called Planck scales (which has spatial resolutions at Planck Lengths, and temporal activities operating in Planck Time).

Sophia: which is or is not the quantum foam?

herb: butt it is more than that my dear (the goddess) Sophia, for is there not the necessary climbing of (length and time) scales that would then bring us to events we see with our eyes, hear with our ears, and measure with our machines?

Sophia: and all of that is driven, if you will my dear herb, by actions of the reality flux, which is the power source ... and, now, it seems to be a lot lot more

herb: my dear (the goddess) Sophia, a lot more indeed, because when things of the reality flux are not in physical Universe, lets agree to put them 'in'[50] a named abode, Negative Universe,

---
[47] a term used by the great shaman Buckminster Fuller.
[48] a term used by the shaman Richard Rorty.
[49] thalamus really wants a better word than 'in' at this moment, because the use of the word 'in' seems to imply that Negative Universe has spatial extent, which who would k-now if that were the case or not
[50] herb, as thalamus before him, wants a better word for 'in' at this moment

and the nature of this non-spatio-temporal abode may have many features that along our journey in these caves (and beyond) may in the end even allow us to think of as the 'real' 'things,' while the reflections in physical Universe could be called mere shadows.

Fate: similar to those shadows on the cave wall which are registered by human (and you k-now who you are) visual systems, but have no particular essence about them, but rather, when understood correctly, make one aware of other essences.

thalamus: other essences from which the shadows result and indeed depend.

the Data of universe

> the interlocutors laugh and laugh as turns are taken making a show of shadow figures on a wall, while all along each interlocker k-nows it is a show that has been endowed and allowed by (the goddess) Fate

herb: lets consider now in more detail the different levels used by science to reason about processes in (physical) Universe.

Giordano: are we to include all 'things' considered by the sciences of physics, chemistry, and biology my dear herb?

herb: most of the 'things' covered by these sciences my dear Giordano; and we should continue on from our discussion of the reality flux and first consider processes at lower levels, which means smaller spatial scales and faster temporal scales, and then we see that these (lower level) processes have influence on physical events at higher levels.

thalamus: is this idea that the lower level events have influence on (or produce?) the higher level events similar to the old ideas of 'emergent computation,' where interactions of the lowest level give rise to higher level organized behavior?

herb: yes, somewhat, my dear thalamus, and if for now we go along with this type of (emergent behavior) thinking, we should consider the following levels of emergent behavior in our discussions:

Science assumes that different particles and forces combine to form certain primitive 'stable' aggregate systems, which we refer to as the chemical elements, $H - U$. Then these 'primitive' aggregate systems combine in certain ways to form molecules, compounds, and polymers, for example. Another important level of possible emergent behavior is the coordinated interaction of larger aggregate systems and forces to produce the 'things' involved in biology.

Giordano: and along this line of thinking my dear herb, should not the ubiquitous carbon atom, and the mysterious hydrogen bond, both be considered important emergent behaviors?

herb: indeed my dear Giordano, as both the carbon atom and the hydrogen bond are key participants in the highest level of localized emergent behavior yet found, the phenomena of intelligent life.

Glia: and what about the so-called Standard Model of particle physics my dear herb, is that part of the particles and forces that produce primitive stable aggregate systems that you began with?

herb: the Standard Model of Particle Physics places 'particles' into three (3) categories: Leptons, Hadrons, and Gauge Bosons. Leptons contain the electron, muon, and tau particles, and each of these has an associated neutrino, and all six of these particles have an antiparticle. The hadrons are all built from quarks, and are separated into two types: Mesons (which act as bosons), and Baryons (which act as fermions). There are six different quarks used to build hadrons, and each quark has an antiparticle ... and this is part of 'the story' (i.e., the narration) of the standard model.

Glia: and is there a difference between bosons and fermions? Because if there is not, then maybe we should call them the same thing, and thus be free to talk of other things.

herb: indeed my dear Glia, and the standard idea here is that particles with a bosonic essence can share the same quantum state, while fermions can not share the same (exact) quantum state; with the best example being, that since electrons are fermions, electrons need form shells around the nucleus in the atomics, where each location in the shell(s) has a different quantum numbering.

Glia: so all the electrons can not be crowded directly in, say, the first shell around the nucleus?

herb: right.

Joe: also, you might add my dear herb, that all fermions are 1/2 – spin particles and are thus subject to the Pauli Exclusion Principle (thus the need for the many 'shells'), and bosons all have integer spin values, and thus are not subject to said exclusion principle.

thalamus: and is it not also assumed my dear herb, that besides this characteristic of spin, much of what you are calling the data of the Universe (computation) typically has other certain fundamental characteristics, including mass, charge, isospin, color, along with that 'spin' thing.

carriers Of force

> *the interlocutors seem to have wandered into a hallway, or avenue of sorts, where not only do the walls contain geometrical drawings and designs, but the walls themselves seem to emanate a glow, and have an almost crystalline, emerald-like appearance and texture. herb continues in his overview of the Standard Model of Particle Physics*

herb: so while we have briefly been introduced to Leptons and hadrons, I will now describe the category of 'particles' k-nown as Gauge Bosons. These gauge bosons are postulated to be the particles that make up the interaction fields for the fundamental forces of Universe (as computation), and are thus viewed as the carriers of (the) force(s).

Sophia: and what is a (the!) 'force' my dear herb?

herb: physical laws governing the time-dependant behavior of the data of Universe (computation) all involve forces (which in turn can produce acceleration in accordance to Newton's law of motion). Modern science currently tracks four fundamental forces (covering gravitational interactions, electromagnetic interactions, weak interactions, and strong interactions), where each force is supposed to be mediated by a gauge boson(s), which are also k-nown as the quanta of the interaction field.

Sophia: can you give a quick example my dear herb?

herb: of course my dear (the goddess) Sophia. For example, the electromagnetic force is believed to be mediated by photons, so then the photon is considered a gauge boson, the carrier of the electromagnetic force.

Giordano: and what about the force of gravity, does this force have a gauge boson also?

herb: it is supposed that the so-called graviton is the gauge boson for gravitational forces my dear Giordano, but as we move forward in our discussions, you will find that the phenomena of gravity may have an altogether different character than what could be accomplished by a graviton alone.

Joe: which actually would make sense my dear herb, with all the dark this and dark that[51] that has been introduced to 'modern' science to explain the large-scale organization of the Universe.

the Code of universe

> *a seating area is arrived upon by the interlocutors, who are simply famished, and lucky for them Joe has that cart tagging along, which contains enough nourishment to provide a nice meal for all. The next day after a morning snack, the interlocutors continue their investigations (into both the physical and mental realms)*

herb: in our time-domain simulations that we are using for our analogy, the 'code' is analogous to physical laws in Universe (as computation). And physical laws governing time-dependent behavior of the 'data' of Universe (as computation) all involve 'forces.'

---

[51] Joe is referring to the hypothetical dark matter and the hypothetical dark energy that has been inserted into the scientific lexicon as a stop gap measure of sorts to explain away many of the anomalies associated the the large-scale structure of the Universe (as we k-now it) that are not explainable by gravitational theory alone.

So then the code of the Universe (as computation) can be viewed as the stuff that calculates, for example, the electromagnetic interaction between data items with 'charge,' which is presumed by the Standard Model to be mediated by the photon; and thus the code of Universe (computation) is here managing the photons.

The weak interaction that acts between data items classified as leptons and quarks, which is presumed to be mediated by the $W^{\pm}$ and the Z^0 gauge bosons, needs also to be implemented in the code of Universe (computation), as does the strong interaction between data items classified as quarks, which is presumed to be mediated by 'gluons.'

thalamus: and at what level does the code of Universe (computation) then operate my dear herb? At the level of gauge bosons?

herb: to think about at what level the code of Universe (computation) interacts with the data of Universe (as computation) my dear thalamus, consider the facts that data of Universe (computation) is embedded in a type of substrate called space-time, and that all code and data must have access to the reality flux.

Giordano: so then the code of Universe (as computation) must use the reality flux to control the gauge bosons that are constantly being required for the various, and constant, interactions between data of Universe (computation)?

herb: very near, most likely, right on right on my dear Giordano, because, as modern science sees it, for example (to use electromagnetics and photons again), for data with a characteristic of charge, the reality flux supports the activity that propagates interaction photons between data items. It is then the aggregate behavior of the interaction photons that is the field phenomena we can describe using the Maxwell Equations, and measure with our machinery.

some Subtleties

> *as the interlocutors begin to exit the cave system at the suggestion of (the goddess) Sophia, herb is anxious to add a bit more to this code/data discussion as it relates to the overall analogy*

herb: as we exit this marvelous, apparently ancient organized enclosure, I need point to a couple subtle issues associated with the analogy we are using between time-domain simulations and the way of the Universe (as computation).

Can not each of us see that the actual code of Universe (computation) needs to 'reside' 'in' that abode we have labeled Negative Universe, for the code does need access, as Giordano suggested, to the control of all 'things' of the reality flux in order to handle the data properly.

Giordano: butt how does this relate to the way of time-domain simulations my dear herb?

herb: nice my dear Giordano, for in a time-domain simulation, while the code of the simulation only seems to reside in memory, it is then typically an interpretation of each machine language instruction (of the code) by a smaller set of programs embedded in the computer which directly controls the electrical signaling that in turn controls all aspects of the hardware operations in the processor. These embedded programs are called the 'microcode' of the processor.

thalamus: and in this way the microcode in the processor acts as a type of 'reality flux'?

herb: very nice 'analogy' my dear thalamus, and this could lend credence to the assertion that the code of Universe (as computation) needs 'reside' in Negative Universe, wherefrom the code activates the needed involvement of the reality flux (which as we should recall is just a part of the quantum vacuum mechanism). So this code / reality flux coordination is very similar to how the machine language instructions in the code (in a computer) activates the appropriate microcode content in the processor to control the machinery (of the computer) that in turn controls the flow of that mysterious fluid type magical substance called electricity.

thalamus: so is that then the complete analogy between the way of time-domain simulations (run on a computer) and the way of Universe as computation?

herb: that is about it, and I think we have seen that these tWo ways are indeed very similar as far as each of the stories go. And so now would be a good time to mention a peculiar category error / mistake that is made by many a modern scientist when they investigate the code of Universe (as computation) using the quantum mechanics. For you see, many a quantum physicist will mistake the quantum entities used in the calculations for the real 'things' in Universe (computation) rather than realizing that quantum mechanics is at best just a description of the workings of the machinery that supports Universe (as computation).

Giordano: a map and territory category error then, so that you suggest my dear herb that quantum mechanics in some way only describes / measures actions of the substrate, that aether-thing we discussed earlier.

herb: quite possibly my dear Giordano; but those quantum descriptions may become more distant from any actual ontological goings on of Universe (as computation) processes as we move deeper into this overall discussion, and consider other possible computational scenarios for Universe.

an Elephant in the physicists room

> *the interlocutors now find themselves somewhat wandering through a forest grove, and as it is, each is also wandering within their own mental forest, each not quite sure of anything, and hopefully the divine would k-now why. Fittingly, the sky and the surrounding turn dark amid a time of the day when the light should shine, evidently that same damn battle again, which may or may not have always been going on*

herb: so we have discussed the standard model and the data of Universe (computation), and the physical laws and the code of Universe (computation), and thus we should now also be advised that modern science wants all physical laws to be 'causal' laws, meaning that only causes preceding in time can induce effects that come later in time.

Giordano: butt my dear herb was not this challenged long ago by the shaman Feynman and his colleagues when they used the 'advanced solution' of electromagnetic problems to show that results of electromagnetic interactions nOw could be the result (at least mathematically) of signals of the absorbers from the future?

herb: indeed my dear Giordano, since in general, for propagating electromagnetic waves, the mathematics presents the possibility of both the so-called 'retarded solution,' which involves waves propagating forward in time, and, the other solution is as you mentioned, the so-called 'advanced solution,' which involves waves propagating backwards in time.

Glia: ah, the future causing the past; how non-causal

AlBe: and quaint

herb: it gets even stickier when we consider how (or if?) 'Mind' intervenes onto the physical Universe (as computation). "For example, dig this: I turn my head, I raise a finger, and then take a wink; here my mental processes seem to dictate (or steer) some actions in the machine that is me. The signals from my brain propagate through my spinal cord and fan out in my body to effect movements in the appropriate muscles in a causal manner to be sure. But what of the initial "thought" wherein I (apparently?) decided to make this gesture?"[52]

Sophia: butt maybe you are mistaken my dear herb, and maybe 'Mind' does not have any causal effect at all (on things in the physical Universe), and it is just a sort of mental motion picture show, or as they say, the mental life is just epiphenomenal.[53]

---

[52] from Ontological musings on how nature computes.
[53] according to the Wiki: "Epiphenomenalism is a mind-body philosophy marked by the belief that basic physical events (sense organs, neural impulses, and muscle contractions) are causal with respect to mental events (thought, consciousness, and cognition). Mental events are viewed as completely dependent on physical functions and, as such, have no independent existence or causal efficacy; it is a mere appearance."

Fate: and if that is the case my dear (the goddess) Sophia, then all the thoughts and actions taken and/or given to humans (and you k-now who you are) are not really theirs at all, but just part of the calculations done by the code of Universe (as computation).

Sophia: and this would be the case only if 'Mind' is epiphenomenal; which I suspect our dear herb will be disagreeing with post haste.

herb: my dear (the goddess) Fate, we will soon see that all should not be lost for us humans (and you k-now who you are), and even though the time it takes, for example, for the signals to travel throughout the typical human body seems to prohibit 'Mind' intervening onto the physical, we may only have to make a few tweeks to get to where it is still possible for humans (and you k-now who you are) to have the possibility of free will.

thalamus: was it not the shaman Libet who ran many experiments that showed the signal lag from brains as they fan-ed out into the body (and vis-a-versa)?

herb: indeed my dear thalamus, and taken at the so-called 'face value,' this simple signal lag time would appear as evidence that thinking can not in 'real time' actually manage the human body in a casual manner.

One solution to this conundrum was brought forth by the shaman Hameroff and his colleague the shaman Penrose when they supposed that a quantum link exists between the microtubules (in brain neurons) and something 'beyond' the physical Universe, and then they further supposed that 'thoughts' that need interact with, say, the body, are actually 'sent backwards in time' some nominal amount in order to provide the needed causal involvement of 'Mind.'

Giordano: and this process that the shaman Penrose used, it needed to access what he referred to as a 'Platonic World,' which is some type of abode[54] that contain those extra things beyond the physical that are needed for mental activity.

herb: right my dear Giordano, as the shaman Penrose was convinced that there are certain parts of mental activity that can not be a result of just physical interactions.

---

[54] the shaman Penrose actually conjectured that the Platonic World was all the 'cubies' at the bottom of space-time geometry.

Giordano: it was a sort of Godel's Theorem[55] type proof that the shaman Penrose presented concerning the non-computability of 'Mind.'

herb: and while not all scientists were convinced by the shaman Penrose's arguments, I can tell you now that this Platonic World-idea, that extra thing outside the physical that the shaman Penrose requires, should be considered very related to the type of abode we have been calling Negative Universe. And in this way we can surmise that 'Mind' 'resides' in Negative Universe, just as the code of Universe (computation) is required to do.

Sophia: so in this way 'Mind' can accomplish some computational steering of Universe (computation) from outside-the-system.

herb: indeed my dear (the goddess) Sophia, this appears to be exactly what is happening ontologically, and this relationship should both remind us of the classic Cartesian Dualism of Mind and Matter[56] and 'signal' (hehe) to us that we should not worry too much that 'thoughts' have to travel back-in-time a bit to get the body to act 'in time' to, say, catch a baseball.

on The nature of gravity

the light begins to shine, and the divine willing, this is what 'it' often does. Butt (the goddess) Fate wants to cause some trouble

Fate: are there any more important 'things' that need 'reside' in that non-spatio-temporal abode we have been calling Negative Universe? I mean, anything besides that little ole 'thing' you are calling 'Mind'?

---

[55] Godel produced a mathematical theorem that showed that all truths about a logical system can not be produced by a logical system itself... something else would be needed. Thus, there is not a complete closed-logical system possible.

[56] the French mathematician, scientist, and philosopher Rene Descartes supposed that res extensia (matter) was in space-time and exhibited traits of locality and causality, while res cogitans (mind) was not in space-time, and thus exhibited traits such as non-locality, and was essentially non-causal – all as told by B.J. Hiley in Non-commutative Geometry, the Bohm Interpretation, and the Mind-Matter Relationship.

herb: the great shaman Bucky spoke of 'Gravity' as "an instantaneous most economical interrelationship of all energy events," and to be instantaneous we must suppose that gravity possesses some non-spatio-temporal attributes, and thus 'Gravity' is a candidate for something that could also 'reside' in Negative Universe.

Sophia: and in this way both 'Gravity' and 'Mind' could be computational steering mechanisms for what we have been calling universe As computation!

and Now what

> *the interlocutors arrive at a place they seem to have been before, where a nice fire pit location stands out, along with a stack or three of wood. A fish in a nearby river jumps out of the water to take a wink before entering back along its course of destiny*

herb: eye would like us all now to be able to separate some of our concerns as we investigate further the physics of Universe as we find it, including the way that the 'Mind' intervenes on the physical (if in fact it does), and this whole Negative Universe, reality flux, physical Universe trinity we seem to be playing with.

Fate: and what is to become of these interesting topics my dear herb?

herb: we should make separate times to speak of these topics, and give them each their proper due.

Giordano: and which should come first? The topic of 'Mind'!

herb: we will save deeper discussions on the nature of 'Mind' and brain for another time and place, as for tomorrow I have in mind (hehe) to begin discussions of another ontological viewpoint of how nature may be computing. We will utilize geometry as somewhat of a first principle, along with some very interesting conjectures that we can attribute to the shaman Arthur Young, the great shaman Bucky, and the shamans Sheldrake and Bohm.

Giordano: so we are to discuss in the detail the computational Cosmography?

herb: tomorrow, yes, but now, after a short poem I have here about that mysterious fluid type magical substance called electricity (or 'light'), we should bring on the band for tonights festivities.

a Dance

the sense of another's presence has arisen from surroundings in which we reside. sight is not yet possible for it has yet to be defined.

how to acknowledge the sensation is beyond my grasp, but before we can help it something has accomplished the task.

is that another, or is it just we. are we separate from the surroundings, or are the surroundings me.

another sense has consumed me. it is not the same as we. we cannot help but to ponder the mechanism of me

before The physical

> *the next sun rise has occurred, but actually the sun more or less stayed where it was (sit!), and it was because the Earth was rotating about an axis that the interlocutors were blessed with another so-called 'sunrise,' while the fire still smoldered from the party the night before, and herb attempts to find the best way to start a discussion of another way of the physical (and not-so physical) world*

herb: if we all together want to step back and ponder the possibility of another basis for the physical Universe, I mean to say, something other than the code / data analogy that we can create by using the modern Standard Model of particle physics, the gauge field theory of forces, and the obvious analogy we get from the ways of computation, I submit we would have to consider the idea of a type of pre-space Pleroma / substrate upon which all 'things' would be fashioned.

Giordano: so then you are looking for a type of 'material' out of which, say, an electron would be formed?

herb: not only out of what an electron is formed my dear Giordano, but a substrate upon which the quantum foam ideal is fashioned, and for which the reality flux-mechanism can interact with, to provide both 'space,' the 'things' within the 'space,' and even 'time' itself.

Giordano: so you are concerned then with the mechanisms used by the Demiurge to create this lowest realm in all of god's creation?

herb: in one way yes my dear Giordano, in that according to the very reliable Gnostic teachings of old[57], even though this lower realm was in fact fashioned in a non-authorized act, it presents many challenges to the understandings of humans (and you k-now who you are); and then of course there is the possibility that humans (and you k-now who you are) will never be able to understand 'reality.'

thalamus: what is the big deal, if I may ask my dear herb, if the Universe was constructed 'illegally,' if you will, because as you intimated at the end there, it seems pretty cool how it all seems to work.

Sophia: the problem my dear thalamus, if I may my dear herb, is that in the construction of the lower physical realm, this Demiurge also as a matter of course entrapped immortal souls from the higher realms (those 'places' hiding away 'in' what herb has been calling Negative Universe[58]).

Butt irrespective of the way and why of the construction, it may behoove the interlocutors to inquire further into herb's line of thinking about some pre-space / pre-matter type thing-y (principle) that the physical Universe may in fact have been built with; while also keeping in mind the apparent connections between what we call physical and the needed, and most likely – of course – the a priori, non-physical.

---

[57] herb here probably means the stories related by the great shaman Simon, and many stories probably found in Alexandria, or in older Egyptian-times (reliable enemies of the sons of Beliel all); as compared to the newer, bastardized version of gnosticism used by so many cults, and also used here and there by that cabal of maniac magicians.

[58] Sophia wants a better term for 'in' when describing 'occupation' in such an abode

herb: and as I asked of us all yesterday, I would like to delay further discussion on 'Mind' (and That immortal soul thing!) until another time, and keep going, now during this lovely divine morning, into a deep ontological foundational basis for what we call the physical.

Glia: and as we work towards this so-called ontological basis, will we get some glimpses of that ole 'hint parade'?

little K-nots of 'light'

(the goddess) Fate has asked herb to explain in simplest terms (please herb!) what is this computational Cosmography thing-y. (the goddess) Sophia seems to have channeled the divine spirit through her flute, Giordano continues to square the circles, while thalamus and Glia dance two the divine tunes

herb: the great shaman Bucky conjectured that all 'things' propagate at that one speed limit[59] of c , so that 'matter' was, by necessity, simply 'K-nots' of light, whose constituent pieces (photons?) operated at this speed limit. The shaman Arthur Young takes this further, and while also surmising that 'particles' are also built from an aspect of that mysterious fluid type magical substance called electricity, namely *light*[60], and then surmises still further when he explains how when 'light' gets trapped into 'particles' it loses a 'degree of freedom,' and then as particles combine into 'atomics,' another degree of freedom is lost, until finally when the atomics group

---

[59] so this then was the great contribution of Einstein when he wrote up the special theory of relativity, because as it is, Einstein used the dilation equations of shaman Lorentz and the shaman Pioncare to do all the calculations; but it was the conjecture that irrespective of the (non-accelerating) reference frame, the speed of light in that reference frame was always c - even if the reference frame was traveling with a velocity of $0.9c$ with respect to another reference frame. Einstein also used these same dilation equations to show that Maxwell's Equations are invariant under these types of transformations, as it was k-nown that the Maxwell Equations were not invariant under the older Galilean transformations... at least that is how the fable goes.

[60] here *light* encompasses any frequency of electromagnetic radiation, and even though 'visible light' for humans (and you k-now who you are) is only a small frequency range (between infrared and ultraviolet), lets just call any frequency 'light'

to form 'things' in the 'molecular realm,' that last degree of freedom (of 'light' from which all this is fabricated) is lost.

Now then, at the molecular realm, of which the great shaman Bucky was fond of fawning over, when he called for witness to those 'beautiful molecular structures,' we have this most symmetrical realm (undoubtedly due to some underlying principle – which we get to very soon), as agreed to also by the shaman Arthur Young, who suggested this beautiful symmetry came at a cost of the loss of all degrees of freedom (of the 'participant light').

So we see now how 'light' goes through this 'fall' (in the shaman Arthur Young's terminology) as it loses some degree of freedom to both create and participate in each realm, and then after the molecular realm, the shaman Arthur Young proposes that the 'light' 'rises' and gains back degrees of freedom as the next realms are manifested. To wit: molecules are patterned into 'things' of the plant kingdom, and here a degree of freedom is gained – as plants have that one single ability of outward growth.

Sophia: and so here then, if I may my dear herb, we have the molecules which make up the plant, which are composed of 'light,' creating scenarios where 'light' in its most 'free' form (sense) is literally eaten by plants (that ole photosynthesis thing-y), and then also as part of this consumption of 'light', still other 'beautiful molecular structures' are constructed by and in the plant.

herb: and this dance has been described by you very well my dear (the goddess) Sophia; but we should add then that these new molecular structures are the foundation upon which the next kingdom in 'lights' continued 'rise' (after the 'fall') is dependent, that is, the animal kingdom depends on these newest style of molecular structures (sugars, et. al.) to exhibit now in this realm two degrees of freedom, as we know the growth of animals has two directions possible.

Fate: tell us my dear herb, are there set patterns so that only certain types of trappings can be 'formed' of that mysterious fluid type magical substance called electricity – in its basic form of 'light,' when it groups into particles, atomics, molecules, oh my

herb: my dear (the goddess) Fate, maybe the shaman Sheldrake can help us here, when he describes the ideas of 'morphic resonance fields,' which are things that are most likely resident in that abode we have called Negative Universe. The shaman Sheldrake proposes that once a specific pattern has ever been created, then that pattern (of K-noting) is then available everywhere and everywhen thereafter.

Giordano: is that similar to the shaman Bohm's suppositions about a background implicate order (that he used in his days in an effort to bring some ontological meaning to quantum mechanics)?

herb: indeed it is my dear Giordano; butt we need step back a bit to (the goddess) Fate's query, because it contained another aspect as it concerned What is allowed.

on How nature computes

> *(the goddess) Fate realizes – and amongst all entities k-nown or unk-nown, who or what could k-now better, that somethings just happen the way their destiny dictates; butt now it happens that (the goddess) Sophia wants to help, possibly out of necessity*

Sophia: my dear herb, you have set the table very well (and, hmmm... I am a bit hungry), but please now discourse with us about those geometrical principles that (you must be about to tell us) govern this patterning of 'light' during the 'fall.'

herb: indeed, and thank you my dear (the goddess) Sophia, for it is well past due that we discuss the synergetic geometry of the great shaman Bucky, and how this fashions the principles of the computational Cosmography, which in its simplest form is an elaborate dance of beautiful geometric forms (with choreography most surely influenced by the divine spirit by 'way of' Negative Universe)

thalamus: and please go way back, if I could insist my dear herb, and give us as much background as possible, as we may now desire to understand why 'light' has certain destinies, and how it could be that 'light' also has the capacity to respond to necessity.

herb: as you wish my dear thalamus, and as it was, it was an early assumption by the great shaman Bucky that provided the key to this background upon which I will now expound. For you see, it was the great shaman Bucky's contention that the tetrahedron was the most important geometrical arrangement we can find in Universe, and it is at the same time the simplest structurally stable space enclosing form possible.

Glia: so then the great shaman did not 'believe' that everything was built from point particles or two-dimensional 'things'?

herb: my dear Glia, no, the great shaman Bucky intuited as the great shaman Planck did also, and thought of the smallest 'things' in Universe (as computation) as quantized units of 'stuff.' Maybe, it could be said, that the great shaman Bucky went a step further than the great shaman Planck, and insisted that this quantization of 'things' was of a space-enclosing nature, or better yet, all 'stuff' was really a 'system'[61].

Now pay attention now, and let me k-now if the following makes sense. The tetrahedron is constructed using four equilateral triangles (i.e., triangles with equal length sides), and then the triangles are put together so that each triangle shares an edge with the other three triangles. This produces an arrangement with a total of four (4) vertices and six (6) edges; here check this figure, which show a tetrahedron inside a vector equilibrium grid (which we will discuss shortly):

---

[61] the great shaman Bucky was found of saying that a 'system' had things outside the system, things inside the system, and then there was a part that was the system itself; because technically, here at least, space has yet to be defined.

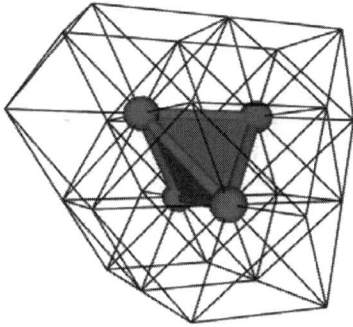

thalamus: and can you not then also produce a vector equilibrium inside the tetrahedron my dear herb?

herb: indeed my dear thalamus, and help me out here as we explain this construction to the other interlocutors. If we connect the mid-points of the six sides of the tetrahedron and connect the now six vertices each to its closet four vertices, we produce an octahedron inside the tetrahedron; as I have drawn here in this figure. The octahedron thus has six vertices, eight faces, and 12 edges[62]. Now, make the mid-point of each of the 12 edges of the octahedron the vertices of a new geometrical construction, the 'vector equilibrium'[63] (VE). The VE has eight triangular faces, six square faces and $(E = 12 + 14 - 2 =)24$ edges.

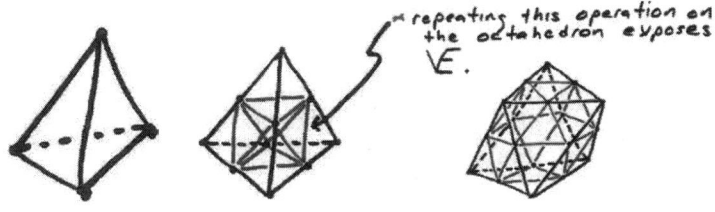

---

[62] which agrees, as it must, with Euler's formula for polyhedron:
$$V + F - 2 = E.$$
[63] also called a 'cuboctahedron' by coexter

And now to an assertion of our dear thalamus, where we see that the VE can also be viewed as constructed internally of eight tetrahedron and six one-half octahedron. To wit: each of the eight exterior triangular faces forms a tetrahedron where the fourth vertice, of each of the eight tetrahedron, is at the center of the VE, and also a fifth vertice (where the other four vertics make up the six square faces of the VE) for each have one-half octahedron is at this same center of the VE.

thalamus: and in this way this whole organization of geometry is contained within itself in this very beautiful manner

Sophia: 'as above, so below' my dear herb; a favorite phrase of our old friend hermes

herb: indeed my dears (the goddess) Sophia and thalamus, because as we see from this construction technique, we find that inside the tetrahedron is found in fact other tetrahedrons, albeit two levels down. Then of course, we can then run that same geometrical construction and find yet more VE inside those eight tets inside the VE found inside the first tet

Glia: and this self-similarity across scales is very similar to the famous mathematical fractals that produce such pretty pictures.

Fate: and when we look closely at any part of a fractal, it seems like the whole of the fractal resides in each tiny piece.

Sophia: 'as above, so below' is the mantra that must also apply to the mechanisms of How nature computes

herb: and now we will go a little further into how we can break apart, say, the tetrahedron, and octahedron to create the so-called A-modules and B-modules. Butt be advised, the great shaman Bucky had many other geometrical constructions within the VE constructed matrix of 'space' itself, which can produce other imaginings of dances of geometrical forms as 'time' is turned on. To be specific, I would invite everyone to investigate the wonderful space-enclosing form called the 'coupler'[64], which by itself is also a space-filler.

---

[64] the coupler relates to the geometry of a rhombic dodecahedron (RD), which is the dual of a VE. The RD has twelve diamond faces, and by itself is a space filler. If each of the 12 faces of an RD are used as vertices along with the body center vertex of two neighboring RD, we get the 6 vertex coupler. Each RD then is composed of 12 1/2-couplers.

thalamus: a space-filler like a cube can be?

herb: right my dear thalamus.

AlBe: and can a tetrahedron be a space-filler my dear herb?

herb: no, my dear AlBe, using only a tetrahedron would leave gaps as you tried to use it to fill space

thalamus: and tell us all what the gaps would appear like my dear herb

herb: as you happen to k-now my dear thalamus, all the gaps in such an attempted construction could be filled with octahedron.

thalamus: so then it is the case that tetrahedrons and octahedrons combined together in just the right way can fill space

herb: right.

Albe: so tell us my dear herb, are A-modules and B-modules space-fillers also?

herb: any one of these modules by themselves my dear AlBe is not a space-filler, butt as it is, the great shaman Bucky discovered that the combination of 2 A-modules and 1 B-module, put together to form a so-called 'mite,' is a space-filler

thalamus: can we then surmise, my dear herb, that any polyhedron that is composed of a ratio of 2 A-modules to 1 B-module is a space-filler?

herb: you may indeed surmise such a thing my dear thalamus.[65] So, the specifics, we break apart a tetrahedron into 24 pieces by establishing a point at the body center of a tet, and then making four polyhedron wherein each uses the vertices of an exterior triangle and the body center vertex. Each of these new polyhedron has $1/4^{th}$ the volume of the original tet, and then each of these new polyhedron are symmetrically separated into 6 polyhedron (by using 6 polyhedra of the triangle face which result when 3 perpendiculars are drawn from the opposite side to each of the 3 vertices), and these now (4 x 6 =) 24 polyhedron are called A-modules. Thus, the volume of an A-module is 1/24th of a tet volume.

thalamus: and a similar construction can be performed on an octahedron where the eight internal polyhedron use a body center vertex and each of the eight triangular faces. Then these

---

[65] and thus it now surprises no one to say that the coupler is composed of 8 mites, consistent with the 2 A- to 1 B-module ratio

8 polyhedron are broken into 6 symmetrical pieces (as was done before inside the tet). Butt here however, each of these (8 x 6 =) 48 new pieces are in fact composed of 1 A and 1 B module, producing then 48 A and 48 B-modules inside an octahedron, and here is a picture created by the shaman R. Hawkins which shows these modules inside the tet and octa

herb: and then finally, we must say that to fill space, we need an arrangement of two tetrahedrons for every one octahedron, giving a total of 96 A-modules and 48 B-modules, revealing that 2-to-1 ratio that has been mentioned.[66]

the Light ~~is~~ in the playground

herb lights up, and at the same 'time' contemplates how best to layout the mechanisms of the inter-play

Giordano: so it is then, my dear herb, that 'form' forms the basis of patterns that we find at both the smallest scales and at the largest scales?

Sophia: 'as above, so below.' (And there must be a way to remember that melody from so long ago that the great shaman Pythagoras taught me that went along with this beautiful saying)

---

[66] thus, if a tet has a volume 1, a coupler will have this same volume of 1 (as it has 24 modules also), and then an octahedron has a volume of 4 - given the same length side as a tet.

Joe: it was said, my dear (the goddess) Sophia, that Pythagoras could play a tune to induce any mood he wanted into anyone

Sophia: and right you are my dear Joe, butt being a theurgist, he would not do so against the will of another, or four his own nefarious purposes

AlBe: it was also said that he could talk to animals

Sophia: and write you are my dear AlBe

Fate: butt herb, with all respects to the great shaman Pythagoras

herb: who was reported to believe that all things are numbers

Fate: indeed, and while that is true (in at least two ways germane to the moment), can we get back to Giordano's query, witch I think I think, was concerned with the patterning of 'light.'

herb: and with great thanks to the divine, and your insistent prodding my dear (the goddess) Fate, I shall now, with the interlocutors permission, embark on some mathematical details of that mysterious fluid type magical substance called electricity, which are, in my humble opinion, needed right a b o u t nOw to point out many fine subtleties of our situation.

I will use the vector equilibrium (VE) as the cell for a calculation of the vector calculus curl, for you see, our recent mathematical and physics history gave us these beautiful *Maxwell's Equations*[67]:

$$\frac{\partial \mathbf{E}}{\partial t} = c^2 \nabla \times \mathbf{B},$$
$$\frac{\partial \mathbf{B}}{\partial t} = -\nabla \times \mathbf{E},$$

---

[67] originally Maxwell couched his unification of the mathematics of electric fields and magnetic fields in the mathematics of quaternions, which at that time was a recent development due to shaman Hamilton. These equations were then updated by *The Maxwellians* - which is a great book by the shaman Hunt describing the efforts of the shamans Fitzgerald, Lodge, and Heaviside (among others), who converted the equations into the more 'modern' vector calculus formulations; with Heaviside doing most of the heavy lifting.

where c is the speed of electromagnetic propagation (or *speed of light*), **B** the magnetic flux density, and **E** the electric field (shown here in SI units for free space propagation). Here then, to be specific, I am interested in calculating $\nabla \times \mathbf{B}$ and $\nabla \times \mathbf{E}$, which in turn must equate to the rate of change with respect to time of the electric field (when the curl is multiplied by c^2), and the (negative) rate of change with respect to time of the magnetic flux density, respectively, as shown in the coupled set of partial differential equations that are the Maxwell Equations.

Further, I will propose, after showing a technique for calculating the so-called vector curl (e.g., $\nabla \times \mathbf{E}$), that this technique can give us insight into the 'how' of lights interaction with the playground that is space-time.

Giordano: so, if I may my dear herb, you are proposing that light is based on a type of 'form,' and then that 'form' is to be related to the 'form' of space-time.

Sophia: 'as above, so below'

> *is the beautiful tune that the goddess continues to play, and play, and play*

herb: and right you both are, for you see, I will try to convince us all, here, nOw, that the great shaman Bucky was correct when he claimed that the VE 'is' in no uncertain terms the *geometry of nature*.

Fate: so then this, in what remains of our mathematical adventure, here, nOw, is to be (finally!) the answer to Giordano's query about the patterning of 'light'?

herb: so I start with the VE cell and place six vectors inside, with the tail of each vector at the center of the VE, and the head of the vectors pointing to six (6) of the twelve (12) exterior nodes, as shown in this figure[68].

---

[68] And it should be noted that those six vectors inside the VE, are also the vectors of a tetrahedron, here just moved around so that the three (3) vertices of the tet on the horizontal plane are moved together (to the center of the VE), which forces the fourth vertex of the tetrahedron to separate (and those three vectors that met at the fourth are now pointing in the directions of $\mathbf{e}_3$, $\mathbf{e}_4$, and $\mathbf{e}_5$) as shown in the figure. Further, the four *planes* of the VE that we shall see shortly correspond in orientation to the four faces of the tetrahedron... cue (the goddess) Sophia's tune: 'as above, so below'

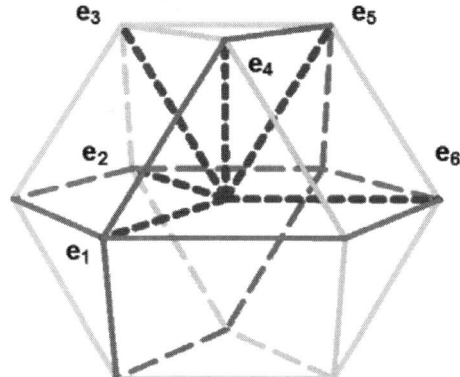

The basic VE cell has twelve vertices around the center node. I use a six-tuple to reference all of these aforementioned twelve points, with each element of the tuple representing a displacement in the corresponding unit vector direction. The center of the VE cell will be designated to have an address of (0,0,0,0,0,0), and the vertice designated with the address of, for example, (1,0,0,0,0,0), represents the point at the tip of the basis vector $\mathbf{e}_1$. A shorthand notation for the vertice at (1,0,0,0,0,0) is [+1]. (That is, I will use [-6] to represent the vertice at (0,0,0,0,0,-1).)

For definitiveness, I will use two vector fields: $\mathbf{S}$ and $\mathbf{T}$; which by construction are related by the equation $\mathbf{S} - \nabla \times \mathbf{T} = 0$ (which is to be evaluated at $\mathbf{S}[0]$). The vectors $\mathbf{S}$ and $\mathbf{T}$ are written is terms of the IVM basis:

$$\mathbf{S} = S_1\mathbf{e}_1 + S_2\mathbf{e}_2 + S_3\mathbf{e}_3 + S_4\mathbf{e}_4 + S_5\mathbf{e}_5 + S_6\mathbf{e}_6,$$
$$\mathbf{T} = T_1\mathbf{e}_1 + T_2\mathbf{e}_2 + T_3\mathbf{e}_3 + T_4\mathbf{e}_4 + T_5\mathbf{e}_5 + T_6\mathbf{e}_6.$$

I start the development of this omni-directional curl operator by using Stoke's Theorem to change the curl operation of $\mathbf{S} = \nabla \times \mathbf{T}$ to

$$\mathbf{S} = \lim_{dA \to 0} (1/dA)[\mathbf{n} \oint \mathbf{T} \cdot d\mathbf{l}],$$

where *dA* is the area enclosed by the contour of integration, and **n** is the unit vector in the direction which makes the right-hand side of the equation take on its maximum value. We evaluate separate contours integrals around the four hexagonal planes of the VE cell. (The four hexagonal planes of the VE cell – shown in figure – are for definitiveness labeled the **a**-plane, the **b**-plane, the **c**-plane, and the **d**-plane.) The aforementioned contours on the four hexagonal planes of the VE cell include only the exterior points of the VE cell, with each plane containing a unique set of six points. The exterior vertices on each of the hexagonal planes (written here in the aforementioned shorthand notation) that make up the four contours are (with the vertices in this order):

a-plane: [+1] [-2] [+6] [-1] [+2] [-6],
b-plane: [+1] [+4] [+5] [-1] [-4] [-5],
c-plane: [-2] [+4] [+3] [+2] [-4] [-3],
d-plane: [-5] [-3] [+6] [+5] [+3] [-6].

For each of the four contours to be evaluated (one for each of the VEs four hexagonal planes), note that the left-hand side of the contour integral equation involves three components of the **S** vector in the IVM basis. For example, to evaluate the contour integral around the **a**-plane (shown colored in magenta in the figure), I take the dot product of the resulting contour with the unit vector normal to the **a**-plane, $\mathbf{a}_a = (\mathbf{e}_2 \times \mathbf{e}_1)/|\mathbf{e}_2 \times \mathbf{e}_1|$; thus

$$(S_3\mathbf{e}_3 + S_4\mathbf{e}_4 + S_5\mathbf{e}_5) \cdot \mathbf{a}_a = \frac{1}{dA}[\oint \mathbf{T} \cdot d\mathbf{l}]_a.$$

For definiteness, the IVM basis vectors can be expressed in Cartesian coordinates (in accordance with a Cartesian orientation that puts two IVM grids inside the Cartesian grid - as it is done with salt crystals), to wit:

$$e_1 = \frac{1}{\sqrt{2}}(\mathbf{a}_x - \mathbf{a}_z),$$

$$e_2 = \frac{1}{\sqrt{2}}(\mathbf{a}_z - \mathbf{a}_y),$$

$$e_3 = \frac{1}{\sqrt{2}}(\mathbf{a}_x + \mathbf{a}_z),$$

$$e_4 = \frac{1}{\sqrt{2}}(\mathbf{a}_x + \mathbf{a}_y),$$

$$e_5 = \frac{1}{\sqrt{2}}(\mathbf{a}_y + \mathbf{a}_z),$$

$$e_6 = \frac{1}{\sqrt{2}}(\mathbf{a}_y - \mathbf{a}_x),$$

then

$$e_3 \cdot \mathbf{a}_a = e_4 \cdot \mathbf{a}_a = e_5 \cdot \mathbf{a}_a = \frac{\sqrt{2}}{\sqrt{3}},$$

which yields

$$(S_3 + S_4 + S_5) = \frac{\sqrt{3}}{\sqrt{2}\, dA}[\oint \mathbf{T} \cdot d\mathbf{l}]_a,$$

$$= a',$$

which defines the *plane variable a'*, that is,

$$a' = \frac{\sqrt{3}}{\sqrt{2}\, dA}[\oint \mathbf{T} \cdot d\mathbf{l}]_a.$$

In a similar fashion, the evaluation of the contour around the **b**-plane results in the following equation:

$$(S_2 + S_3 - S_6) = b', \text{ where } b' = \frac{\sqrt{3}}{\sqrt{2}\, dA}[\oint \mathbf{T} \cdot d\mathbf{l}]_b.$$

The evaluation of the contour around the **c**-plane results in the following equation:

$$(S_1 - S_5 - S_6) = c', \text{ where } c' = \frac{\sqrt{3}}{\sqrt{2}\ dA} [\oint \mathbf{T} \cdot d\mathbf{l}]_c.$$

The evaluation of the contour around the **d**-plane results in the following equation:

$$(S_1 - S_2 + S_4) = d', \text{ where } d' = \frac{\sqrt{3}}{\sqrt{2}\ dA} [\oint \mathbf{T} \cdot d\mathbf{l}]_d.$$

Evidently, as the equations from the four contours attest to,

$$a' - b' + c' - d' = 0,$$

which we will call *the principle Of no local holding patterns*.

Furthermore, a solution showing how to calculate **S** [0] can now be written:

$$S_1 = (c' + d')/4,$$
$$S_2 = (b' - d')/4,$$
$$S_3 = (a' + b')/4,$$
$$S_4 = (a' + d')/4,$$
$$S_5 = (a' - c')/4,$$
$$S_6 = (-b' - c')/4,$$

which shows that each vector component of **S** is dependent on the contours around two separate hexagonal planes.

The solution for S_3, uses the a and b hexagonal contours:

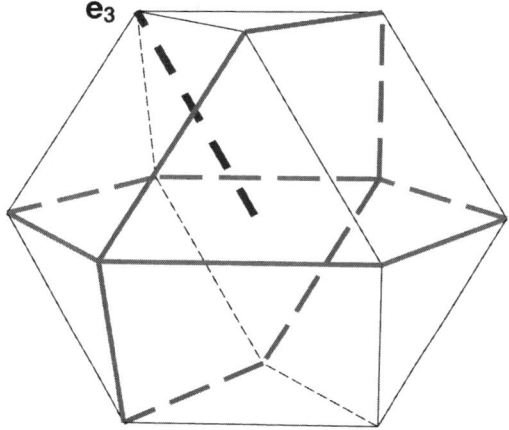

whence the electric field as tetrahedron, with three equal valued components forming 3 of the six sides of a tetrahedron (from a propagating solution):

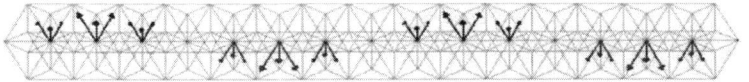

principle Of no local holding patterns

the interlocutors do not have long now before they are to part their separate ways for a score or two of days, or a moon or two or three, butt always, it will be as (the goddess) Fate would have it

Giordano: so then herb, we see how light in a sense can be viewed in terms of four hexagonal planes that inter-play to more or less 'hop' from place to place in space, and/or from time to time in time; butt how does that compare to the ideas of the shaman Arthur Young concerning light forming into K-nots as it begins 'a fall' into matter, and then experiences 'a rise' to participate in that phenomena we call 'life'?

herb: so then with those four planes you mention my dear Giordano, consider again the four plane variables a', b', c', and d' which for 'light' to act in an electromagnetic way needs to have the relationship that

$$a' - b' + c' - d' = 0,$$

which I have called the 'principle Of no local holding patterns.' So as long as the 'light', or electromagnetics radiation, a version of that mysterious fluid type magical substance called electricity

Sophia: that last name is my favorite non de plume

herb: indeed; so, as long as the mysterious fluid type magical substance called electricity adhers to the principle Of no local holding patterns, and as the hexagonal planes of its essence interact, then the 'light' can be described by these so-called Maxwell Equations; butt when we do have local holding patterns (of the 'light'), we necessarily will call the light something else

thalamus: words like 'particles', 'atoms', 'molecules', and the like

herb: right. For example, if we trace the path of each plane variable as the order is described in the principle Of no local holding patterns we get the flows as modeled in this figure. Note that all of the eight triangles on the outer VE has a somewhat 'circular' flow, while each of the six square faces has 2 source nodes and 2 sink nodes (and each node is part of two triangles and two square faces, and a source node on one of the two square faces is a sink node on the other square face).

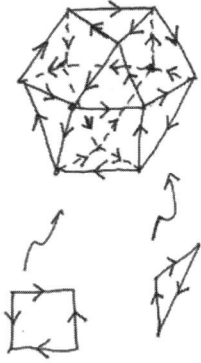

Then, lets now suppose the principle Of no local patterns does nOt hold, then a node (or organized sets of nodes) will not have a zero-sum flow, and now the VE can go through what the great shaman Bucky called a jitterbug process and, say, 'fold' into an octahedron, where the 24 outer edges of the VE now double-bind into the 12 edges of an octahedron (and this may in fact explain 'current flow' when that mysterious fluid type magical substance called electricity more-or-less has a 'phase change' as 'it' flows on the skin of wires or within the realms of crystals).

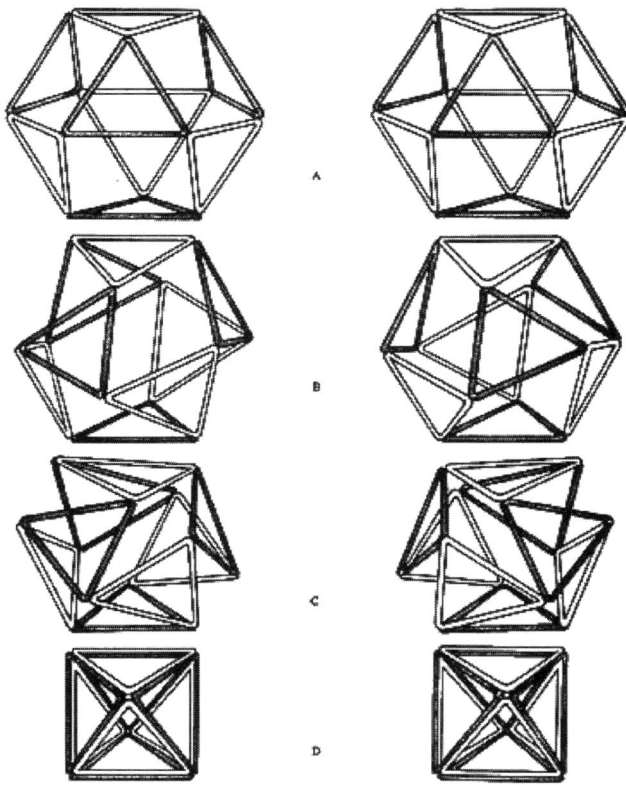

Still more interesting however is when more extreme variations of the principle Of no holding patterns occurs, and the VE goes past the octahedron phase and contorts via the jitterbug process into a tetrahedron, which has 6 edges, each of which has 4 of the original 24 edges of the VE… and in this way we have a principle on how K-notting can be done to "form" so-called particles… and the vision of the shaman Arthur Young is explained as is the great shaman Bucky's contention that all phenomena operates at the speed of light, whereupon some phenomena involving little K-nots of 'light' are just more K-notted up than others.

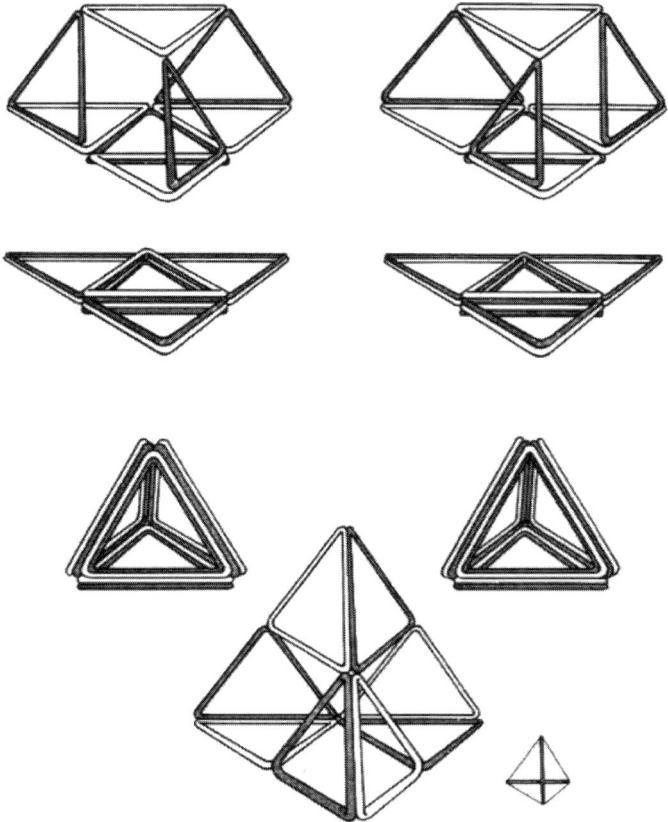

AlBe: so this is then what happens in the internals of the many Sun's, where little K-nots of light are being folded into particles and these 'particles' are then arranged with others to make the atomics involving protons, neutrons, and electrons (oh my!)

herb: and these atomics become part of our celestial resources – a paraphrase of the great shaman Bucky - when the many Sun's send them (i.e., the atomics) forth throughout the vastness

Sophia: so now then my dear herb, we have covered a lot of physics, and you have convinced us interlocutors that it is all about a flow of that mysterious fluid type magical substance called electricity, which can be viewed as a flow of number also, as suggested by the great shaman Pythagoras, butt what of that Elephant in the physicist's room, what of the physics of mental processes?

herb: and right you are, this simply must be our next discussion topic, butt if memory serves, (the goddess) Fate has suggested we save that until our next get-together in a moon or two

AlBe: in two moons it is then, at which time we can party and discourse 'on Mind'

herb: butt before we close this part of the Act which has been 'on Physics', eye would insist we take one final overview of this dance that is the physical part of Universe

Giordano: the dance involving that mysterious fluid type magical substance called electricity

herb: and this dance then allows for many options at every point in time and space, for if we use the VE as the model, can not you all see then that we have a crossing of 6 ways (or maybe actually 12 ways if we take into account both the positive and negative directions of each vector of the tetrahedron) at the center of the VE; and thus these evidently are the 'ways' in which the 'influence' of light - in the physical - can take effect on things along their course of destiny in association with necessity.

It would then be 'forms' and intentions from Negative Universe that can act as a bit of 'code' to steer that mysterious fluid type magical substance called electricity in all of its manifestations (including electromagnetic with all its freedoms, or the manifestations that occur along its fall and rise as it gives up and then again regains these freedoms), which is to say that this elaborate flow of number is managed in such a way is it not to produce those mysterious physics and biologies that occur in the membranes of life forms, and even more mysteriously, make up the actions of the ATP that coordinates communication at a somewhat meta-level between all the major biologies that make up the human brain and nervous system, providing in a very mysterious way the vehicle for a journey of the immortal soul within this lowest realm in all of god's creation

Sophia: as then 'light' in all its manifestations navigates the 12 ways, and every 'point-instance' of the materialists' space-time-Pleroma, did you not just say my dear herb, that there needs be some control exerted, as then did you say this control emanates with intentions and 'forms' from that abode of Negative Universe, by way of a common connection to all 'things', and thus the physical Universe is not just physical.

herb: my dear (the goddess) Sophia, eye did indeed imply such a thought / thing. The ancient theosophists took a more ancient term to describe a similar stance, in that the *akâsâ*, for example, which from old times represented the spiritual life principle, a sort of omnipotent agent involved - by necessity - in every phenomena / 'thing'.

Giordano: and then this *akâsâ* in a sense is guiding this intricate web of number along its flow, fall and rise

herb: it would seem to be the most plausible explanation for how the Pleroma, Mind, and matter seem to be utilizing techniques that may in fact have an ontological basis my dear Giordano

> *as (the goddess) Fate would have it - and does not she always get her way - it will have been three moons before the interlocutors join together again*

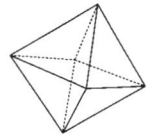

INNER-MISSION

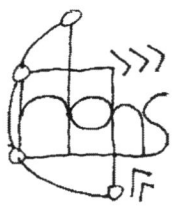

on Mind

Love is the law, love under will.
- Fra. P.

> *after the third moon, and a big one it was, a super moon if you will, full it was at its moment of perigee (if we allow Luna to be in possession of such a thing as a perigee), the interlocutors have got the band back together at a beautiful location that was suggested by thalamus and Glia*

herb: so we all k-now it is quite popular to think and speak of Universe as a computation

Giordano: which we have been doing a lot of, if you do not 'Mind' (hehe) the interruption my dear herb

herb: not at all my dear Giordano, and I agree, we have also been doing these popular things, and as we all k-now, this is possible because computation has indeed changed philosophy drastically

thalamus: and not just from the point of view that it is easier to run thought experiments and / or do mathematical simulations

herb: correct, the idea of computation has literally opened up whole new vistas on how to reason about the different patternings, histories, and futures of the many faces of the K-nots of 'light'

Sophia: and on these points we must all surely agree my dear herb

the Fable of the simulation argument

> *some of the interlocutors are still arriving, and all with a smile when they greet the others and when they greet their surroundings (hello surroundings), which includes a shimmering lake, many lovely trails, nice camp sites, a fire pit, and a small amphitheater for the band, oh my*

herb: we should begin this glorious day by summarizing the Fable of the simulation argument, a famous ole story possibly (in a Galactic way of things[69]), butt brought to us in most recent times by the shaman Bostrom. Within this Fable, the shaman Bostrom also takes the view of Universe as a sort of computation, butt he then steps much further and begins to provide scenarios where 'Mind' also is to be considered a type of computation

Giordano: and is not it obvious that 'Mind' is a type of computation my dear herb?

herb: it may indeed seem obvious my dear Giordano, butt many obvious things are not what they appear to be[70]. To wit, as the great shaman Bucky was oft to say, it is still all so fashionable to say (tosatosa) that the Sun 'comes up' each morning, even when we k-now now, do we not, that this is not what is happening during a sunrise

Giordano: and your point is well made my dear herb, as it is the fact that the Earth is rotating about an axis that we are blessed with a day and night, combined with the fact that the Earth revolves around the Sun, and not the other way around[71]

Sophia: and so my dear herb, if 'Mind' is not a computation, then what is 'it'?

herb: my dear (the goddess) Sophia, as is well k-nown, the great shaman Descartes proposed that 'Mind' (his *res cogitans*) was something with an essence not confined in space and time; meaning that 'Mind' is most likely resident in that abode we have named Negative Universe. And-d as we shall sea in due course, if 'Mind' requires residence in the abode we have named Negative

---

[69] the shaman Bostrom himself proposed that this topic he addresses in the simulation argument would most likely be available to any advanced society throughout the expanse that is physical Universe.

[70] in the Harry Potter fables of old, there were moments where Harry was convinced that Serious Black was a son of Beliel (a sOb), as the propaganda made it appear obvious that this was the case. Later, Harry would learn that he was wrong about this 'obvious' issue, and would realize that Serious Black was in fact a son of the Serpent, for ever at battle with those sOb's.

[71] which Giordano remembers well caused many many to suffer horrendous pain and misery when in olden days it was blasphemous not to follow the dogma of an Earth-centric universe

Universe, that may indeed imply that 'Mind' is not a computation that can be done only with physical quantities

thalamus: butt, with respect my dear herb, have not many disputed that claim, and even proposed to put 'Mind' into computers?[72]

herb: indeed my dear thalamus, so it seems appropriate to discuss the Fable of the simulation argument, and bear witness to just such a proposition, and all that such a proposition entails

Fate: butt does not this Fable contain much more than just the supposition that 'Mind' is nothing more than a result of brain activity, and thus then one must suppose, 'Mind' **is** a type of computation, where here the brain would be the computer that the 'Mind' program is running on?

herb: right you are my dear (the goddess) Fate, as the Fable of the simulation argument is much deeper that the simple question of whether or not 'Mind' **is** simply a result of brain activity

Glia: so how are we to start my dear herb?

herb: the shaman Bostrom supposed that one, and only one, of a possible 3 options could be true:

> • *the fraction of human-level civilizations that reach a 'posthuman' stage is very close to zero*
> • *the fraction of 'posthuman' civilizations that are interested in running ancestor-simulations is very close to zero*
> • *the fraction of all people with our kind of experiences who are living in a simulation is very close to one*

for you see, can you not, that if human societies never reach a 'posthuman' status (which, I probably need add, if societies can go posthuman, it may mean that these societies become presumably free of the concerns of war and disease, and also become armed with unimaginable levels of computing power); then the

---

[72] thalamus of course here is referring to the 'Brain builders dream'

truth of things is in option 1), and then we do not have to worry about option 3) being the truth of things

Glia: and this may seem reasonable to assume that all technological societies eventually blow themselves up (in some way) before the societies can come to terms with how to live with advanced technology, where for example, there becomes little need for much of human labor that was the way for humans (and you k-now who you are) to 'make a living,' I mean, with robots and all, what is there for the common wo/man to do with their life then?

Sophia: a very good point my dear Glia, for it is not really the case that a human (and you k-now who you are) wo/man needs only just food / sleep and entertainment to live a full life; as there should also be a way, should there not, I mean ideally, for that innate spirituality and drive to help their fellow wo/man to have an outlet

Joe: butt maybe we do not even need to go that far to find faults from the ideal my dear (the goddess) Sophia, but rather simply notice that human (and you k-now who you are) ways of organizing society in the 'recorded history'[73] simply does not keep up when real technological progress kicks in

AlBe: 'damn the torpedoes' my dear interlocutors, as maybe eye refuse to accept that every time a society of human-type entities (anywhere resident in this vast physical Universe) comes to a 'point in time' where they can advance past the simple, competitive, it must be 'you or me' (as the great shaman Bucky was oft to say) stage, that all (these societies) must perish!

Giordano: and AlBe has a valid point here, the divine willing, in that it does not seem the Fates would damn every society, on the verge of advancing to a competent computational usage, into oblivion

Fate: it would not seem a 'logical' thing to do

Sophia: so then my dear herb, if option 1) is not always the case, the divine willing, what are we to think about then?

---

[73] which our dear Joe k-nows to be in fact mostly blasphemy; a distortion 'recorded history' mostly is, a tool of the sons of Beliel (those sOb's) to manage and mangle the 'Minds' of wo/man

herb and Giordano seem to be in a disagreement about something. (the goddess) Sophia informs the other interlocutors that the discussion may concern whether or not now is the time to bring up that issue of the sons of Beliel (those sOb's), and what this cabal of maniac magicians impact could be on the veracity of option 1) in the Fable of the simulation argument. (the goddess) Fate comments that maybe all human-type societies live in a similar land of shadows during some part of their development

a Functionalist masquerade

tim has drawn (and not with a pencil) his boat to the shore of the shimmering lake, which reminds some of the interlocutors about a summer long ago on a mysterious Ludington shore

herb: so then if option 1) is not the case, and a society moves to a posthuman stage, what are 'they' (i.e., the posthumans) to do with their time and the virtually unlimited computing resources.

Joe: they would probably play games all the time?

Sophia: and really complicated games one would imagine, eh my dear Joe?

herb: and it is true my dear Joe and my dear (the goddess) Sophia that many a philosopher supposed just such an answer to a similar type situation

Giordano: right then my dear herb, you must be speaking of the 'Jupiter Brains' scenario

herb: indeed my dear Giordano, where it was supposed that if all the mass of a giant planet like Jupiter could be placed into an organization such that the entire mass (of the planet) would have thinking capability, then these 'super brains', or 'Jupiter Brains', would most likely pass their 'time' playing involved games with other 'Jupiter Brains'

Glia: like the shaman Hesse's Glass Bead game?

herb: one could only suppose; butt eye am not all that interested in supposing the specifics of what a posthuman would do, but rather, eye would like to discuss the shaman Bostrom's next suppositions as it pertained to option 2) (in the Fable of the simulation argument)

thalamus: and, what then, my dear herb, did the shaman Bostrom suppose?

herb: the shaman Bostrom supposed that some posthuman societies might be interested in running (on their vast computing resources), what he called, 'ancestor simulations.'

AlBe: and why would a posthuman be interested in an ancestor simulation?

thalamus: maybe the posthumans would want to get a better close-up look at how they themselves had come to be, by creating, in no uncertain terms one can not help but suppose, a type of 'way back machine'

Giordano: butt my dear herb, does not this idea of an ancestor simulation then suppose, that 'Mind' can somehow be both created and also made resident in a computer?

herb: and right you are my dear Giordano, it certainly assumes such things. For the latter, the shaman Bostrom is very upfront in his suppositions, and in fact discusses in detail the idea of substrate-independence, which is a way of supposing that 'Mind' (whatever 'Mind' is) can be run on (or be resident in) some 'things' other than brains.

Giordano: so then the computer was supposed to be just another substrate upon which 'Mind' could reside

herb: and that is part of the rub eye think my dear Giordano, for you sea, here the shaman Bostrom has taken a very functionalist point-of-view as far as 'Mind' is concerned, to wit: the shaman Bostrom has taken the assumption that 'Mind' is just a result of brain activities

thalamus: so then, if 'Mind' is supposed to be an emergent result of brain activity, then with proper posthuman programming, it would seem reasonable that from the interactions of a substrate that is a computer, then a 'Mind' could also emerge

Sophia: and hence, the Brain builders dream

eyes Wide shut

herb dips a pinky toe, picks up a mask, while thalamus rolls

herb: so you see then, do we not, why eye would call such a simulation, a Functionalist masquerade?

AlBe: would you say masquerade, because this type of computation, in your 'Mind' (hehe) my dear herb, seems nothing more than a type of motion picture show?

Giordano: and we have discussed this before, this idea of whether 'Mind' is just epiphenomenal, meaning, as our dear AlBe suggests, more like a show or a play

thalamus: and if eye recall, we heard you, my dear herb, extol the virtues of a dualist view of how 'Mind' intervenes on this place we call the Universe

Fate: or rather than a Universe, it could be just a posthuman computation scenario, where you, and all the interlocutors are just ancestor simulations being run by those from your future, so that they can understand better, who they are

herb: and as (the goddess) Fate would have it, this leads us to option 3) of the Fable of the simulation argument, and if the interlocutors would allow it, eye suggest we put aside the question of functionalism versus dualism for right now

u R living in a computer simulation

nOw, as all the interlocutors recall, option 3) of the Fable of the simulation argument demands that we all need be living in some type of posthuman simulation; Giordano, not quite convinced yet of the logic, suggests a short break, with maybe a hike or boat ride, and thalamus suggests a smoke, while herb, shockingly, asks where the beer is, because everyone of the ancestor simulations have been told to be thirsty, and they might as well enjoy the show

Fate: so the Fable of the simulation argument has brought us to option 3) as the current topic of discussion, four if we deny that all societies collapse before reaching posthuman stage

AlBe: witch would seem a divine possibility that all societies are not to be damned

Fate: and then we also considered the case where some of the posthuman societies do decide to run ancestor simulations

thalamus: because they may be curious how it indeed happened that they escaped from 'Zeus's tyranny'

Fate: and irrespective of these rationales, desires, or even childish dreams, can you explain my dear herb why then with the falsity of option 1) and option 2), would the interlocutors have to accept the shaman Bostrom's option 3)

herb: indeed my dear (the goddess) Fate, and it is here that the shaman Bostrom uses large number probability arguments, for you see, can you not, that we must consider other possible civilizations on other possible planets in these type of calculations

Giordano: and here then is the effect of large numbers, for there are more than billions of billions of star systems as we look outwards from the Earth, and that is within even the smallest amount of solid angle of the night sky, and then if even a very small percent of these possible civilizations in these other star systems 'pass the test' (lets call it) of option 1), then there is set to be billions of billions of posthuman societies about

Glia: right my dear Giordano, and then the shaman Bostrom must make the same type of probability arguments about these billions of billions of posthuman societies, and if option 2) of the Fable of the simulation argument is false, then, because of large numbers, there would be billions of billions of posthuman societies running ancestor simulations.

AlBe: and from these discussions, and our understanding of the shaman Bostrom's use of large numbers, which seems appropriate for the vastness of space

Sophia: for surely humans (and you k-now who you are) would not be the only intelligent race

AlBe: then why would option 3) have to be correct, for could it not be possible, the divine willing, that we, nOw, be part of an original race?

herb: my dear AlBe, here again we should be cognizant of these large numbers that are being used in the Fable of the simulation argument, for can you not see, if there are billions of billions of posthuman societies, wherein many are running 'ancestor simulations,' then on average any of the remaining societies that are not yet posthuman, have a larger probability of being one of these simulated societies[74] rather than a society in an original Universe

Sophia: if of course there is such a thing as an original Universe

herb: which brings us to a very important place, for before we dispense with this Fable of the simulation argument, we should ponder further on how some of these posthumans would behave.

tim: and what of the shaman Bostrom, my dear herb, did he offer any conclusions as to witch option, 1), 2), or 3), that he preferred?

herb: the shaman Bostrom, at the end, my dear tim, did contend, that

> *"Unless we are now living in a simulations, our descendants will almost certainly never run an ancestor-simulation."*

thalamus: and so, if we can conjecture, and with our k-nowledge of the sons of Beliel (those sOb's), what say you herb, are we living in a simulation, and are our (rr) mental thoughts just a program, running on a type of machine?

herb: here then, we must go back, to an assumption the shaman Bostrom made, and remember eye called it a Functionalist masquerade

---

[74] 'being run from some posthumans couch,' or so opined a writer in a story that was printed in that archaic news source, the 'New York Times,' before their deception and tyranny became so obvious, that no one would continue to purchase their outlandish blasphemy

what Is a posthuman to compute?

> *the interlocutors break for dinner, they enjoy sounds produced by the band, in the amphitheater, oh my was it grand. After the show, (the goddess) Sophia plays the divine, and she prods herb to continue, his investigation, into, a posthuman 'Mind'*

herb: so we will take as a given that option 1) of the Fable of the simulation argument is false, and agree with my dear AlBe, and desire the divine be will'ing not too damn two extinction every last one of the, eye suppose, what would be real human (and you k-now who you are) societies. So we come now to the second option of this Fable, and it should not surprise anyone, that eye am to modify this disjunction, summarily cast out the possibility, that the art of being human (and you k-now who you are) is simply a Functionalist masquerade

Sophia: and further, if I may, my dear herb, if not simply a Functionalist masquerade, then if a posthuman desired, or planned, to put 'Mind' into a simulation of a wo/man, then these posthuman programmers would be, up against a very large moral quandary

herb: and, my dear (the goddess) Sophia, what would that bee?

Sophia: these posthuman programmers, or posthuman simulators (pHs), if we can all agree, would be acting just as our famous Gnostic demiurge, in that the pHs would be entrapping 'Mind' into something of their own creation, to be specific, 'Mind' would be entrapped into their Universe simulation

Glia: and this entrapment done by the pHs demiurge, would in fact be required, because as us interlocutors seem to agree, 'Mind' is not 'resident' in the physical, butt participates from behind a curtain, if you will, from a mightily different realm

Giordano: and this then is a result of the dualist thinking, that herb has previously ranted about; butt what then to think, my dear (the goddess) Sophia, about these, can eye say, nasty pHs demiurge's who would do such a thing

AlBe: as building an ancestor simulation, involving, both you and me; is that what you mean?

herb: we are all then agreeing with (the goddess) Sophia's insight, that if a pHs were to build such simulations, it would involve more than just computing our 'Minds,' and eye would go much further and suggest they would have to simulate the whole damn thing

mechanisms Four the simulation argument

> *herb has convinced the interlocutors that they must dig deeper into option 3), and think of the whole kitten (meow) kabuttal (sp?) as something a pHs might build. The interlocutors are to start again the next day, soon after the rising of the Sun, when, or about, (the goddess) Fate has informed, a new interlocker will be arriving, that mysterious scientist, dream master, and artist, Amy G. Dala, whom the interlocutors all call amygdala for short*

herb: eye would like us all to consider the case that we are in a pHs simulation, which at the bottom of the computation is a seething dance at the Planck scale of all the K-nots of 'light,' which then leads us to an important question, which will require research, into the actual ways, that a pHs could 'code' such a place

thalamus: so then, my dear herb, if eye understand you write, we are to imagine the physical Universe as just such a place, built by, what (the goddess) Sophia aptly named, a type of pHs demiurge

herb: my dear thalamus. this is exactly right.

amygdala: hello all you interlocutors, and is not this a delight, rumor has it you are having discussion on the Fable of the simulation argument, witch eye find quite intriguing, butt at the same time it presents me with one thorny thought ... how it is then that a pHs could control such a place

herb: welcome my dear amygdala, about time you showed

up, and am I correct in assuming that Giordano has briefed you on most of what has been discussed?

amygdala: yes, my dear herb, as eye understand it, we are discussing option 3), and nOw back to my thorny question, which could be asked, I suppose, by either you or me

herb: the question, plainly stated then, if you allow me, is

> how would our supposed pHs implement algorithmic control of this Universe?

Giordano: indeed my dear herb, and grand welcomes my dear amygdala, as this insight is indeed thorny, and how could it be, that a pHs could control you and me?

Sophia: so then my dear herb, am eye write in assuming, that as we proceed, you are to propose mechanisms Four the simulation argument option 3)?

herb: and in order to do that, my dear (the goddess) Sophia, we need a few philosophical asides, for do you not see, that if this physical Universe is indeed a simulation, then *all* 'things' should be in doubt, including all our beloved physics, which would now just be 'code' for the machinery

Giordano: and what about time?

thalamus: and what about space?

AlBe: and if it is all a pHs simulation, how can the divine intervene into such a place?

observation Selection effects and the nature of time

> *the interlocutors are still near the shore of a lake, the Sun shines brightly, oh what a day. (the goddess) Sophia suggests, amongst all this, that the interlocutors need get back underway*

Sophia: oh dear herb, oh tell me, how should we proceed? For it seems we have already discussed physics at length, those K-nots of 'light' participating in all the various realms during the fall into molecules, and during a rise when it all seems to become scaffolding for 'life.' If it is a pHs simulation that humans (and you k-now who you are) are living in nOw, then your old view of physics with 'code' and 'data' would, still, seem about right

thalamus: and if a simulation it is, we would not have a need for the so-called 'big bang,' for as herb has explained to me, a causal simulation can be started at any time when using the proper initial conditions

herb: eye think it is important at this time to introduce a new idea, brought to us also by the shaman Bostrom, while he was still in school, his dissertation it seems turned into a book, "Anthropic Bias: Observation selection effects in science and philosophy."

When we learn what a selection effect is, it might make clear, and help us in determining how a pHs could exert algorithmic control, on this big physical Universe simulation that we find ourselves in.

Fate: humans (and you k-now who you are) are subject to selection effects all the time, for if humans (and you k-now who you are) can not see, smell, or hear something, then you would not k-now it is about

Sophia: except for those strange little feelings, that all humans (and you k-now who you are) have, those 'goosebumps,' or tingly hairs, or a glimpse from the shadows in the corner of an eye

Fate: for as has been discussed before, every human (and you k-now who you are) has a need for a reality-tunnel, for the simple fact that the totality of senses are not overcome

herb: and these are examples of selection effects to be sure, as these are the most interesting kinds, the selection effects that happen when an observer can not be about

Giordano: and this then is the shaman Bostrom's so-called observation selection effects?

herb: exactly my dear Giordano, and eye think these observation selection effects, which make it clear, that if an observer is not present or able, to observe a particular 'thing,' then that 'thing' is in fact not 'observed'

Fate: and, accordingly, my dear herb, you or another may then not k-now that that 'thing' is about

herb: that is exactly the essence of this particular selection effect; but we need take it further and analyze what this means, as it will help us tremendously when we discuss philosophically, the difference between causality and teleology.

To wit: In the physics of our most recent times it was always the hope to put the theory and mathematics into situations where we plainly had a cause followed (in time) by the effect

Giordano: and this is how all physical laws which use forces are setup; for example, apply a force to a mass, and accelerate in time it does

thalamus: but is not cause-and-effect k-nown to be a bit problematic my dear Giordano, for many a mathematical statement can work both ways in time.

herb: lets make sure we are familiar with these two words, as we need them to proceed, the words again are causal and teleological, and this may help us all as we attempt to observe.

Joe: that word teleology; is that similar to the idea of backward causation my dear herb?

thalamus: backward causation is very popular in biology, for we all k-now every tree comes from a seed

herb: so causal processes are those processes that involve cause and effect that progresses 'forward in time'

Joe: which is what we are used to; and can it be any other way?

herb: teleological processes are more related to the those processes whose required end results determine the intermediate actions leading to said results

thalamus: so in a sense, if eye may my (mm) dear herb, teleological processes, like backward causation, seems to involve cause and effect that progresses 'backwards in time'

Glia: there is that famous electromagnetic example, the Wheeler-Feynman absorber theory, where the so-called advanced solution (which proceeds backwards in time) can be used for some calculations

herb: you interpret things well here my dear thalamus, and for you my dear Glia, you are exactly write, for the Green's function that is used to determine the propagating field due to an arbitrary source; this Green's function admits both a retarded solution (forward in time) and an advanced solution (backward in time)

amygdala: and in most cases, it must be admitted my dear herb, the causal viewpoint is arbitrarily adopted.

thalamus: and that is true my dear amygdala, even in biology

nomological features and the enslavement by time

the interlocutors fashion a break, and each has need for some solitude, which may be a plan delivered - somehow - from the future, or just part of the typically absurd

herb: so we can all agree, eye would hope, that it is a general (or major) desire in science to couch ideas about the natural world in terms of causal processes.

Butt this tendency could in fact be simply an observation selection effect. This particular observation selection effect is most likely the result of our apparent immersion in a forward progression of (the so-called) time.

Giordano: due, in no small part, to, as the great shaman Christopher Knowles would call it, Saturn's, or Chronos's enslavement (of humanity), possible, through the use / enforcement of 'time.'

herb: indeed my dear Giordano, the sons of Beliel (those sOb's) would seemingly have it no other way

Giordano: butt the great shaman Christopher Knowles did, eye think, believe, that we could break Saturn's spell, or in his other words, escape Chronos's tyranny.

herb: right, butt we should first understand, a nice 'observation' by the shaman Jan Faye

> "Our ordinary notion of causation does not track any nomological feature of the world. What counts as the cause and the effect depends on the observer's projection of his or her temporal sense onto the world."

AlBe: so, if we reflect appropriately, we can easily admit that we k-now not how it really is... the nature of things could be causal or teleological; if I allow myself, that could easily bring a scare... butt there is one thing eye do ponder at this moment, and eye want you to answer me, how does this relate my dear herb, to our discussion of option 3)?

herb: indeed my dear AlBe, eye am so glad that you asked, because now in this framework, I can easily suppose, that a pHs could use, both 'Mind' (and 'Gravity') for teleological control mechanisms for this physical Universe we call our world.

amygdala: and for sure we await your ideas about the inclusion of 'Gravity,' butt my dear herb, this is what the shaman Penrose and the shaman Hammeroff eventually supposed... specifically, they eventually required that thoughts need jump backwards in time some nominal amount, in order to control appropriately the microtubule interactions in neurons that lend to the functioning of humans (and you k-now who you are) in this very funny place

herb: exactly my dear amygdala, they found that that (thth) was required for the possibility, that our experiences are not just ephemeral, butt leave open a chance for a decision by our 'Minds,' that can then intervene, allowing me to, say, hit a baseball, or laugh at a joke.

Giordano: and it was not only the shaman Penrose and the shaman Hammeroff that supposed some needed shenanigans are about, the shaman Herbert had that hidden-variable theory of consciousness that contends that physical events are controlled by consciousness through the laws of quantum mechanics

herb: which fits when we consider the need for a Negative Universe and a reality flux

Giordano: and the shaman Bell's non-locality, and the shaman Bohm's hidden-variables and pilot waves

herb: all acting teleologically, by way of the reality flux acting through the strings of Fates weaved by the totality of the little K-nots, remnants of the original goddess some would say, as options are taken in 'time' along the 12 ways

Giordano: and one more thing, in that that (thth) secretsun blogger, the great shaman Christopher Knowles, seemed to agree with us all, when he claimed through experience and insight that

> *mind is not bound by the strictures of time and space and causality*

Sophia: an ontological observation about time and space and causality (and teleology this seems to be

Glia: a key ontological fact indeed, one that seems to have been osbcured from humanity for how long we can not be sure

thalamus: eye hope all this eventually leads us back, to an in depth discussion, of the Brain builders dream!

a Fuller gravity

> *herb wants to bring everything to the nOw, from both the future and the past, while the interlocutors are still grappling with the Fable of the simulation argument option 3), herb is ready to elucidate a non-spatio-temporal essence of that 'thing' called 'Gravity'*

herb: the great shaman R. Buckminster Fuller talked about 'Gravity' as an instantaneous most economical interrelationship of all energy events.

And this is heavy (hehe) because it supposes does it not, that there is no time lag in a decision by 'Gravity,' four if instantaneous, as the great shaman supposed, then 'Gravity' is different from our other forces we have seen, and thus it is not mediated by a gauge boson, that so-called graviton, as the other forces are supposed, no no, 'Gravity' is more like a web, or a tensional characteristic of this K-nots of 'light'-show.

amygdala: so if 'Gravity' is like that, it most certainly resides, right there 'in'[75] Negative Universe, co-habitating with 'Mind'

Giordano: and thus both can act as teleological prime movers, being programmed by the divine? of maybe a pHs? how do we decide?

herb: butt for sure, this would allow, through the use of 'Gravity' and 'Mind,' some algorithmic control

AlBe: so these then are the mechanisms Four the simulation argument, that you are to propose, butt does that mean u R living in a computer simulation? and so there is no free will?

gnostic Theology

herb pulls out a book, and reads / paraphrases from page forty two

herb: the shaman Hans Jonas in his book, 'The Gnostic Religion,' gave a fantastic synopsis of how some of the ancient Gnostic thought viewed this physical realm. For to answer or even discuss my dear AlBe's query about the absence of free will, we should take an aside in our continued analysis of option 3).

As the shaman Hans Jonas comments:

> "The cardinal feature of gnostic thought is the radical dualism that governs the relation of God and world, and correspondingly that of man and world. The deity is transmundane, its nature alien to that of the universe, which it neither created nor governs and to which it is the complete antithesis: to the divine realm of light, self-contained and remote, the cosmos is opposed as the realm of darkness. The world is the work of lowly powers which though they may mediately be descended from Him do not know the true God and

---

[75] 'amygdala wants a better word than 'in' at this moment

obstruct the knowledge of Him in the cosmos over which they rule. The genesis of these lower powers, the Archons (rulers), and in general that of all the orders of being outside God, including the world itself, is a main theme of gnostic speculation, ..."

And the point eye would make here is that the gnostic Theology supposes that this realm is built by a type of pHs, under another name. Here though, even if there is some simulated attributes that contribute to the presence of 'Mind' in the physical Universe we find ourselves in, gnostic Theology maintains that there is still needed that spark of the divine, and thus 'Mind' needs fall down (if you will) into the Universe from some other realm. Can we see then, that this type of arrangement, where any actual 'Mind,' by necessity, still has a connection to the divine, and this would allow for free will. I think this is quite opposed to the shaman Bostrom's premise concerning the ancestor simulations, which thus are a mere Functionalist masquerade, which would seem to lack a free will component

Giordano: or a connection to the great shaman Plato's Divine Mind, or the shaman Huxley's Mind-at-Large

amygdala: or the shaman Penrose's underlying Platonic reality

herb: who can really be sure, but suffice it to say, that mental activities, and thus 'Mind,' does seem to interact directly with Universe in total through this connection to the divine

AlBe: so then even if this physical realm is built by something other than a true 'god,' you would contend my dear herb, that we can still have, an option of choices, true ones eye mean, a real chance at free will, no matter how absurd that all seems

Sophia: my dear AlBe, if our dear herb is correct, and if a demiurge or pHs built this place, and if 'Mind' is not created by matter interactions as supposed in the shaman Bostrom's Functionalist masquerade, then mental activity must reside in some other abode, allowing both for free will and a connection to the divine

Fate: and this is the moral quandary we mentioned before, a decision must be made by a pHs to trap in its 'program' 'things' that have in their nature something divine, perhaps even an immortal soul

herb: and there could be no other name for this particular blasphemy, we must agree to call it what it is, straight-up sorcery

the Brain builders dream

> *all the interlocutors ponder how could it be, that the shear detail and beauty of such a 'machine,' a physical realm that has signals both Forward and backward in 'time,' a place built from tiny K-nots, exhibiting time and space scales so vast, one can easily get lost, butt they hope nOw that herb can provide, another anchor, and help them out, of this distress, because it seems apparent to all that there are some things about this lower realm that do not really make cents*

herb: eye would, with the interlocutors permission, like to tell a short tale, it is about a computational scientist named Fred, and Fred's brain machine that he presumably built

tim: how did he build it to match his own brain, was it with one of those brain slicing machines?

herb: we can assume that Fred discovered some type of non-destructive Brain mapping technique, so that he was now there, next to his machine, contemplating throwing the switch

thalamus: here is Fred privy to the functioning of 'Mind,' does he k-now it resides in another realm, does he k-now that he possesses an immortal soul, and thus a connection to the divine?

herb: and this is the quandary that this particular Brain builder must face, because Fred's 'Mind' is currently connected to his (physical) brain, and if he were to turn on the machine, then what would his 'Mind' do?

Giordano: for nOw, eye sea, the quandary he is in; for his 'Mind' will now have two choices to which to connect, does it stay connected to the real Fred

thalamus: or does it connect to the new machine?

herb: and if Fred's 'Mind' connects to the machine instead of his brain

Giordano: that would leave Fred (brain-)dead

thalamus: butt his 'Mind' would carry on, connected as it is now to a machine (presumably in Fred's lab).

AlBe: and what would the machine do then when 'Fred' wanted to take a walk?

herb: this points out some of the problems associated with the Brain builders dream

thalamus: right, what would the machine feel, and what would it see?

Giordano: lets assume Fred is clever, and thought this out, for then he would hook up sensors and such so that 'Fred' in the box would have a similar reality-tunnel to that Fred with a brain

Albe: so that the box will walk around with arms and legs?

Joe: and would 'Fred' in a box take a shit once or twice a day, would it / he eat any food, oh my, would he / it be able to enjoy a beer?

herb: most likely the machine would go 'crazy' right soon, for it would be no longer Fred

Giordano: and this would explain why all those 'Mind' machines in the deep dark seckret labs are psychotic as soon as they turn the switch on

thalamus: maybe they should change the programming, to bring it up, and, eye mean, nurture the poor 'Mind' in a box like it is a human (and you k-now who you are) baby-being

amygdala: butt should that even be allowed, if it is at all possible, to entrap 'Mind' in a machine, to be a Gnostic demiurge nOw seems the same as the Brain builders dream

AlBe: and you say this my dear amygdala, because you realize, that the 'Mind' they would put in this machine would not be created by the box, but rather that 'Mind' needs fall down, as we have discussed, from some other realm

herb: and into this 'black Iron prison' that the shaman P.K. Dick supposed himself in

on the flexibility of time and the need for the divine

the next day the interlocutors seem sad, they will soon depart this place, mostly their separate ways, promising, however, that they will get back together, and continue to ponder 'on What is'

(the goddess) Sophia is whispering to herb, insisting he find a way to close the Act on Physics and on Mind. herb, the goddess insists, needs to pull it together, maybe talk about the computation that is done in Negative Universe (as the shaman Tiller has tried to do), and what it is that 'they' need, maybe ask again why materialism is so in vogue, when the sons of Beliel (those sOb's) all in fact believe in astrology, and what about that damn all-seeing eye. Or even what could it be that the secret societies k-now, with the threat of death they keep these seckrets, oh my, they will not tell. And why is real history hidden, and why is sorcery allowed? Why does NASA worship Egyptian gods, and not tell the truth about neighbors here, below, and beyond the curtain of the reality flux, and why is not the nature of time and space studied in detail by all, odd this is, almost absurd, that those sOb's seem to want everyone stupid, or is it enough that they be simply sitting on their couch

herb: we should say our good-byes, I k-now we wish all of the each others well; so eye say be well and be blessed, until we rendezvous again

ACT III

on What is almost A midday sun who Do you believe what Did the great shaman Simon Magus suppose on Being born into this world demiurge / Archons / oh my step Om rocks not people on What is allowed a Legacy of the ss for Before the boy k-nows enough most Dangerous of all undertakings only Matter matters (ekki) an Objective reality? both Forward and backward in time a Jitterbug of space the Secret teaching of all ages on Some theoretical problems with brain emulation wild Grey sweet earl orange it Is a great mystery that Life thing conspiracy Of hidden history a Dark magus of history a Wandering who? god And devil and Wo/Man's relationship to it patton Must die satan Came, satan saw, and a civilization falls into ruins a Sekret history of amerika war Is a racket not So hidden judea Declares war not So federal reserve war Funding rabid Dogs have full sway an Atlantis of plato thoth And poimandres a Great flood the Invader venus crisis Of now war Is peace wag The dog the Order of the day old Time religion a Bit too much to think worthless Eaters babylonian Money magick psychology, Propaganda, and written history lilly Waves and mind hacking on Mind control and Henry Ford's assertion that 'all history is bunk' a Hidden religion?

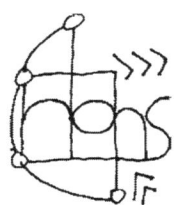

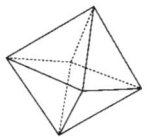

INTER-MISSION

time Fades   time Fades   times Fade (tFtFtF)   a Tale of human existence in the land of shadows, or how wo/man saw light And life (A pOpol vuH) those Divine emanations rules 4 theurgy back 2 ontology light And life false Flags and control and That immortal soul thing and Now a message from Bob – open at the end

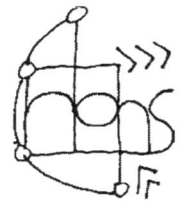

on What is

As always, the path of spirituality is a knife-edge between abysses.
 - Aldous Huxley

herb and other interlocutors are down by the river, and Giordano asks how one would go about explaining to someone the essence of What is

Giordano: through our discussions and personal reflections, we, each of us, has the chance to contemplate deeply about those 'things' that really matter. You know herb, those things that are part of our connection to the Universe.

herb: maybe I k-now my dear Giordano, but please describe more about this connection, or what is meant by those 'things' that really matter.

Giordano: there are so many things that are part of the daily life of a human (and you k-now who they are), butt are there also things about being human (and you k-now who you are) that transcend the day to day?

Sophia: you mean things like ideas about 'god', the Divine, and destiny (and OK, necessity)?

Joe: those things / thoughts that make us smile when a slight breeze of a conception passes by our conscious awareness.

Giordano: these are both to be included, my dear (the goddess) Sophia and my dear Joe; that is, we should include these 'things' into the realm of 'things' that are worth contemplating. I suppose I am interested in what is to be the proper course of instruction for one who wishes to understand things in general, maybe, I suppose, a pre-language type understanding.

herb: my dear Giordano, you wish to discuss the matter of 'What is.' My favorite 'thing' to contemplate, and I wish I had prepared well a discourse that should well and truly put us on a course to understanding, or at least contemplating, correctly, 'What is.'

Joe: should we go over to the amphitheater on the shore, where we can enjoy a smoke and beverage while the discussion begins?

almost A midday sun

> *the interlocutors have been joined by thalamus and Glia, who have been told by herb the previous day that a grand discussion concerning 'on What is' would begin on this very day, at the amphitheater, on or about the 1/6 position of the sun in the sky, that is, before, the midday sun*

herb: to get started properly, we should consider starting with what I will term the 'Pythagorean basics.' As is well k-nown:

> All philosophy is a result of an attempt to explain the nature of Universe, and man's relationship to it.
> - Pythagoras

I dearly love to talk, debate, and ponder 'bout the nature of Universe.

Sophia: which is well k-nown my dear herb.

herb: indeed, and to begin our adventure my dear (the goddess) Sophia, towards contemplating 'What is,' I will start with a diatribe concerning the conspiracy of 'who Do you believe.'

who Do you believe

herb: We can start with a seemingly obvious ontological fact that we are born into this world. The language we begin to speak, and the technology we come to know in early life are very dependent, we should all of us easily agree, on the particular time-space locality whence we entered the / this physical realm (through this particular birth), and the whereabouts where our first 7 or so years (i.e., our first 'set' of molecules) are spent.

Soon after this 7th (or 8-9th) year, as we work on gaining our second 'set' of molecules, up to our 14th (or, say, 17th) year (or so), it would be a hope that each individual also has the divine providence to become inclined to understand both their physical world around them and their dream world that is, evidently, thrust upon them.

I would contend that the start of an individuals spirituality, for a lack of a better term, is quite definitely dependent upon the particulars of their 'Being born into this world' particulars.

Sophia: yes herb, all humans (and you k-now who you are) have those different experiences that you allude to, while admittedly having very little recall of some of the earliest times right after 'Being born into this world.' Butt also, I can not butt feel you are making some assumptions about this world during this early part of your (latest) diatribe.

thalamus: if I may dear herb, and with many thanks to (the goddess) Sophia, you are in fact / or seem to be / maybe / assuming some ontological particulars about the / this world (especially when you stated so frankly that *we should all of us easily agree*). To be blunt, I would hope, and at the same time not be surprised, if you are to include for us, on this splendid day down by the river, an overview of some more of those Gnostic suppositions concerning the / this physical world.

what Did the great shaman Simon Magus suppose

> *herb does not mind really, telling it like it is, and after all, Giordano did ask about 'on What is,' butt the sounds of the water rolling, and the leaves in the wind, it makes it easy to forget how 'Mind' got trapped into this realm*

herb: you know me well dear thalamus, and I must, of course, out of obligation to you and others, necessarily include an outline of the impact of the great Gnostic ideas; brought down to us thanks in part to the great shaman Simon Magus.

Glia: are the Gnostic particulars part of your so-called 'Pythagorean basics,' presumably being concerned with both the 'nature of Universe,' and 'man's relationship to it'?

herb: indeed my dear Glia. The nature of our relationship to the / this physical realm is in fact impacted by the ideas brought forth in the tenants of the classic Gnosticism of the great shaman Simon Magus. For our present purposes, I can make comments on how these aforementioned tenants relate to the issues of 'Being born into this world,' and the conspiracy of 'who Do you believe?'

Sophia: how far back will you delve my dear herb, for as you k-now well, this well is very deep.

herb: eye will not disappoint my dear (the goddess) Sophia, as I will include some historical placement before setting forth the issues / ideas most related to our current charge(s) concerning 'Being born into this world,' and the conspiracy of 'who Do you believe?'

Giordano: should you take care with the inevitable blasphemy so as not to befall my ancestors fate?

herb: let the Divine shine today, let tomorrow bring what it may ...

on Being born into this world

> *the gathered interlocutors are apprehensive as they know well of what Giordano alludes to, and all together decide to follow herb along a hidden forest path as he ventures forth knowing that providence has finally allowed / demanded the interlocutors to go full wyrd*

herb: (lighting a cigarette during a pause on the path) the Gnostic teachings most likely go back to the time before the Great flood, and are most likely born from the ancient battle between the sons of Belial (those sOb's) and their rightly wayward progeny, the sons of the Serpent. Our current incarnations of the teachings however, as aforementioned, are handed down by the great shaman Simon Magus, but only officially came to light in recent historical times with the discovery of ancient scrolls attributed to the Nag Hammidi library and the Dead Sea scrolls. (It would, as we all would agree, also be beneficial if the cabal of maniac magicians' important library, in this day managed by an all too famous ancient blood cult, were to be opened to inspection.)

Sophia: so you think more documents related to Gnosticism are to be found under the Vatican?

herb: (cough cough) as is well known, the beginning of the current edition of the so-called 'church' of the all too famous ancient blood cult was historically due to the great Demon Constantine, and he and his enforcers proudly attempted to genocide all the Gnostic religious groups of those days and eliminate their writings; starting somewhere around 300 AD (as dated by the Gregorian calendar). But what is less well known is that the ancient fire on the library of Alexandria was a False flag terror attack where in fact the greater amount of the contents of said library were confiscated by representatives of the sOb's, and are in actuality still theoretically available to us today.

Joe: what are these teachings, that are germane to 'Being born into this world,' and the conspiracy of 'who Do you believe?', which so frightened the sOb's?

herb: my eternal gratitude my dear Joe, for bringing me back on topic, so I shall nOw relate those aspects of Gnosticism brought most recently to us by the great shaman Simon Magus, that relate to our current topic of 'Being born into this world,' and the conspiracy of 'who Do you believe?'

Glia: should we really be talking about this, here, now?

herb: let the(se) seeking fool(s) continue on their path(s) in safety my dear Glia; butt we should also always allow (alalal) - at all times - for the divine spirit to provide council to all things, along their course of destiny, in association with necessity.

thalamus: are we there yet?

herb: quite right my dear thalamus.

demiurge / Archons / oh my

> *herb leads the interlocutors slightly off the hidden path towards the opening of a large cave, long rumored to connect to a labyrinth that can take one*

> *into a subterranean world of antiquity. AlBe has been waiting for a while, and has set out a meal with enough nourishment for all, complete with some good enough wine and some associated party favors. There are also whispers that later, on or about the 5/6 position of the sun in the sky, that there may or may not be music arriving*

herb: the Gnostic belief is that there are many 'realms' to this Universe, and that the / this physical Universe that is many times postulated as the *only* Universe, is but the lowest realm in all of Divine creation.

We should look at this in much detail, as many issues come to 'light' (as they say). Man, under the Gnostic system, is more than just his physical being.

Sophia: and woman?

herb: yes my dear (the goddess) Sophia, women and men, wo/man... maybe I should say human? (and you k-now who you are.)

Sophia: ok my dear herb. So, as you are saying, 'wo/man' is more than their physical being, but please say more about those other, dare I say, Divine aspects?

herb: indeed my dear (the goddess) Sophia, butt alas, all the ultra-physical is, sadly, not all Divine (as the Gnostics well k-new). The consciousness in wo/man is connected to the fact that each human (and you k-now who you are) particular physical incarnation is associated with an immortal soul / spirit / entity. It is a Gnostic belief that it was a *mistake* to entrap immortal souls in such a physical realm that we find ourselves in. Further, the Gnostic contention is that the / this physical realm was actually built by a sort of lesser god, a so-called 'demiurge.'

AlBe: my dear herb, if there is a lesser god, does this imply a pantheon of gods; and if so, is there a top (dog) 'god'?

herb: according to many ancient tradition my dear AlBe, there are many realms besides this physical realm. At a later time I will speak about how the later science of our most recent (historical) times came to also suppose - although implicitly I mind you - that there is a required something more than the physical, the earlier mentioned (and warned about) non-physical (if you will).

Sophia: why wait my dear herb, can you explain more about the 'non-physical' now?

herb: well, my dear (the goddess) Sophia, I suppose I can give a glimpse by telling you nOw that the sons of Belial (those sOb's) in their so-called 'Kabbala' label the realms above the physical as a *Sephiroth*, a sort of set of imaginable spheres if you will (surrounding, I suppose in a spatial analogy, the physical realm). The sOb's then contend that a 'soul' descends through the Sephiroth as it enters a body in the physical realm. This interpretation, is the same as with many others things that the sOb's take from others; that is, it is a simplification *and* corruption of more ancient ideas[76]; originally due to others with a more Divine interpretation of our reality - as compared to the patriarchal psychopaths that we refer to often as the cabal of maniac magicians (i.e., those sOb's).

Joe: my dear herb, you seem to have gone 'off the rails' a bit as you entered the discussion of the non-physical - if you would pardon my frank analysis. For me, I would certainly want to know if there is any, dare eye say, scientific validity to the idea of non-physical realms.

herb: indeed my dear Joe, and thank you for bringing us back to this important point. So, the Gnostic system also insists on the reality of astral / ethereal beings / entities we can call the *Archons*. (In the latter part of this last societal wave of so-called 'movies' - i.e., story telling through a recorded play of sorts - the 'Smith' character in the Gnostic-themed movie 'The Matrix' could be viewed as a sort of Archon.) Under some systems, these *Archons* are typically soulless non-physical entities in league with a demiurge (and almost certainly in league with representatives of the sons of Belial - those sOb's). Under other speculative systems,

---

[76] see, for example, the great shaman Manly P. Hall's magnus opus *The Secret Teaching of All Ages* for many other interpretations of the ethereal world, co-opted by the sOb's for their – typical – nefarious purposes.

the Archons could even be part of the psyche of the sOb's as they roam the astral / ethereal realms, as part of their tireless work towards the completion of their dark machinations.

Sophia: can others besides the sOb's interact with these Archons?

herb: my dear (the goddess) Sophia, it is most likely the case that others have no choice in the / this physical realm - built by the demiurge(s) - but to be susceptible to influences of the Archons / macrobes / oversouls. There are stories that the constant whispering of the Archonic *locomotive breath* can overcome some individuals in the / this physical realm. Other stories relate how individuals may form pacts (i.e., deals) with them, or even allow the possession of their current physical incarnation (as is most certainly the case with various leader figures associated with the sOb's).

Please note, this situation, associated with the Archonic whisper, has been parodied - as some of you have already intuited - with the little red 'devil' that pops-up on an individual's shoulder, and then tells them to do those dark things that do not respect the Divine instructions associated with civility to others. Often time of course, when the 'devil' pops-up in these parodies, a little white 'angel' also pops-up on the other shoulder to recommend a different choice.

Glia: are there techniques to limit the Archonic influence upon our particular mental state?

herb: yes, my dear Glia. I will try to give a short overview of possible techniques, but we should remember, as we move on, that these issues associated with the Gnostic musings as I relate to you now, the hope is that it provides assistance in understanding issues associated with 'Being born into this world.' After my next statement of this issue, we should move on to discuss 'social norms,' and the related idea of 'What is allowed' to be discussed.

There are (theoretical) techniques that one can use to limit the Archonic influences, and to control the immortal souls current incarnation in the / this 'system' built by a demiurge. These techniques involve the use of divine magic, which can be called theurgy. theurgy is quite different from the sorcerer's magic, in that sorcery involves subverting the will of others, while theurgy eschews such techniques. The simple truth about the sorcery as practiced by the sOb's - flagrantly in violation of the divine instructions associated with civility (to others), is that it involves hidden (or not So hidden) techniques to subvert the will of others.

Sophia: were those rampant advertisements of olden days, and the multiple sources of governments propaganda also to be considered sorcery? Because the use of those messages were surely meant to subvert the will of the peasants in ways hidden to them.

step Om rocks not people

> *after a meal, the band 'theurgist Of atlantis' played some of the ancient popular songs associated with not submitting to the will of the sons of Belial. As a lyre plays softly in the background, and another moon begins to appear, the discussion continues*

herb: I still think it important that we continue to talk about the conspiracy of 'who Do you believe,' but let us now agree to move past the issue of 'on Being born into this world.' Towards this end, I will claim that the social norms of the time(s) associated with an individual souls incarnation into the / this physical world have a huge influence on the early thoughts of an individual. For example, if everyone in the tribe was always naked, then a young member would find it odd to see someone fully clothed. However, in contradistinction to this, if everyone in the tribe was always clothed in public places, then a young member would find it odd to see someone walking around naked. It is simply down to the social norms an individual finds themselves immersed in.

Besides the fact of clothing optional or not, there are other types of social norms that escalate to a larger, more subtle, and potentially dangerous (for the physical body) level.

Sophia: are you talking about religious norms my dear herb?

herb: yes my dear (the goddess) Sophia, and it is not just within religious norms that one can find what can be called *dogma*. In general, we should suppose that dogma is neither good nor bad, but a dogma of the time(s) could be a rock solid belief (on the part of many), and then this particular belief would not admit any real free inquiry unto the merits (or lack of merits) associated with the belief.

Joe: my dear herb, would it properly be called *heresy* to discuss the merits (or lack thereof) of a dogmatic belief?

herb: heresy must be included as a possibility here indeed my dear Joe. The heresy being an idea, or stance / belief opposed to the current dogmatic belief. It turns out to be the case that heresy could be a description of ideas that question norms that deal with religion, and /or individual or group behaviors. Typically however, heresy is associated with the questioning / challenging of religious norms.

on What is allowed

AlBe: why cannot humans (and you k-now who you are) be free to talk about whatever they want to talk about? We mean as long as they are civil to each other of course.

herb: very wise suggestion / observation my dear AlBe. As it was, there were / are places where disagreeing on certain thinking can be punished in fact with a death penalty. In other places, questioning certain historical narratives can get one arrested (as with the famous case of the Canadian academics who were charged, tried, and convicted for offending a certain faction of the sons of Belial - those sOb's).

thalamus: were the academics being mean or threatening to the sOb's?

herb: no, my dear thalamus, they were simply doing academic inquiry (of the highest caliber I might add), which was deemed not to be allowed. How subtle mind control can be then if certain ideas (or historical narratives) are placed off limits to discourse.

Sophia: these are sad stories indeed my dear herb, but maybe there are some things that should not be questioned.

herb: if that be the case my dear (the goddess) Sophia, then one would truly have a difficult time deciding 'who Do you believe.' To wit, if there is only one story (of our past, and our present predicaments), then is it easy to decide whether you believe the story - as told - or do not believe it? Without the ability to hear / read other interpretations of events than those presented in, say, the propaganda of governments and / or the sOb's, how is one to decide / determine which side of the coin you should choose?

a Legacy of the ss

> *as the moon, our darling Luna, reaches her zenith for this particular evening, the group of interlocutors gather near a glorious firepit that has been arranged this particular evening by thalamus. There is still plenty of beverage and party favors to go around, and Joe wants to continue the days investigation into 'on What is'*

Joe: my dear herb, are there other issues, besides those elucidated throughout today, that can impact greatly upon our decision of 'who Do you believe?' (as we continue investigating the nature of 'What is' - as had been suggested early this morning by our dear Giordano).

herb: yes indeed my dear Joe. One of the biggest factors that we can not overlook is concerned with the Legacy of *secret societies* (ss). This is the case because some of these ss groups may have secrets / k-nowledge that may concern:

- how the physical and non-physical realms 'work' / operate
- ancient and / or hidden histories and techniques (including techniques for both theurgy and sorcery, Earth energy manipulations, grid alignments, etc.)
- information about ongoing conspiracies

Sophia: my dear herb, do you mean to say that some of those old guys who sit around a checkerboard floor have real sekrets about the Universe / and all the people in it?

herb: indeed my dear (the goddess) Sophia, it is possible. Butt, how can we be sure about such things... I do not k-now for sure. Because it is the case that those who join an ss are sworn to 'secrecy.' This same type of situation arises with most military and government organizations also, in that they are required to keep mum (sometimes under the threat of death? Think about the fabled-Majestic 12, in that here there may be some sekrets that could benefit humans (and you k-now who you are) if said sekrets were to be made public.)

We can not be sure if these groups are just a mutual *one hand washing the other* style fraternity (of sorts), designed to steal and abuse the others not in the ss, or whether or not there is something involving deeper mysteries.

Giordano: when you say others my dear herb, do you really mean us peasants?

herb: the main point my dear Giordano is that it (i.e., the Legacy of the ss) has restricted assess (four the rest of us not in the ss's) to current - evidently - available k-nowledge for far too long. I hope everyone can understand why this creates a problem for those deciding 'who Do you believe.'

Glia: my dear herb, besides just keeping things / information from us peasants,

> *hehe is heard from almost every peasent present among the interlocutors, which is the entire crowd sans (the goddess) Sophia*

are there other things we should be advised about as it pertains to these ss's?

herb: indeed my dear Glia, as it is not only what they are not telling us that impacts our decisions / thinking concerning 'who Do you believe,' but it is also *what they are* telling us that might do just as much harm. You see, it was the case, for example, that in the heyday of that globally interconnected network, the one called the internet that operated on the mysterious fluid type magical substance called electricity, that many of the so-called websites were actually faked (and *astroturf* was the name given to these type of websites by those who got tired of the false information being deployed).

thalamus: are you talking about the fabled *agent provocateurs*, the well k-nown use of *pysops* (i.e., pyschological operations), that were / are employed by the sons of Belial (those sOb's)?

for Before the boy k-nows enough

herb: there is a great quote my dear thalamus, that I recall from a famous control handbook, brought to us most recently, by sorcerers in a Queen's court. It goes something like

> "... for before the boy knows enough to reject the wrong and choose the right."

From this, we should all try to understand, and hold empathy, for those who do not currently understand the vile influences on our current (and recently past) world, due in no small part to the machinations of the sons of Belial (those sOb's). You see, it may be the case that there are those individuals who have not needed yet (in their current physical incarnations) to become aware of the sOb's.

AlBe: my dear herb, who could possibly not see all the fear and hatred being infused into the / this world by the various propaganda mechanisms?

Sophia: my dear AlBe, maybe deep down, for some individuals, it is just to scary to consider the possibility that the world is not as they have been told.

most Dangerous of all undertakings

> *thalamus puts more wood on the fire, while he and Glia meander among the interlocutors to query whether there is something or another they can get for them... as it seems to them that this discussion will take a while*

herb: thank you my dear (the goddess) Sophia, as I believe you have answered AlBe's query most accurately. This current predicament reminds me of a quote by the shaman Sri Nisargadatta, which goes something like

> The search for reality is the most dangerous of all undertakings, for it destroys the world in which you live.
> - Sri Nisargadatta

AlBe: so does this mean that it is possible that individuals 'tune-out' the influence[77] of the sons of Belial (those sOb's) because they do not want to have to live in a different world than the one they are currently in?

Giordano: my dear AlBe, you have intuited the true nature of herb's and (the goddess) Sophia's comments. As the shaman Huxley noted in his now (in)famous 'Perennial Philosophy,'

> The path of spirituality is a knife-edge between abysses.
> - Aldous Huxley

Thus, if some particular individuals do take the opportunity to study just a little bit, they may be forced to reject the wrong, and choose the right. However, they simply may not be ready yet for their world to change so drastically (and we must realize that their world would change drastically because they now think about everything in the world differently).

---

[77] 'if you do not see the FNORDS, they can't eat you' opined the great shaman RAW

herb: eye think we have now completed 1/2 of the Pythagorean basics we have set out to delineate. I would like now that we as a group move on from the conspiracy of 'who Do you believe,' and venture forth in our dialogue into the conspiracy of 'materialism,' which is part of the second 1/2 of our Pythagorean basics.

only Matter matters (ekki)

> *(the goddess) Sophia has put on warmer clothes and found her guitar. She plays Divine melodies that inspire the interlocutors, who also decide to dress materially properly*

thalamus: my dear herb, can you tell us what you mean when you say 'materialism?'

herb: yes my dear thalamus, it would be my pleasure, as it is very important to understand this term. This idea, that is, materialism, is a favorite modern hammer, used by the sons of Belial (those sOb's), in their effort to build their... well, whatever it is that they (the sOb's) are trying to build.

Glia: is not materialism the belief, or dogma, that physical matter is the only fundamental reality, and that all being and processes, and phenomena can be explained as manifestations, or results of matter?[78]

herb: yes indeed my dear Glia, but more pointedly, materialism is a preoccupation with, or stress upon material rather than intellectual, or spiritual things[79].

Sophia: so can we then all agree that a belief in materialism excludes a belief in the Divine?

herb: thank you my dear (the goddess) Sophia, and my dear Glia for pointing out the real issue here. That is: If one supposes that only matter matters (which is of course not (ekki) true), then there is only the / this physical realm; and the reality of non-physical realms (abodes) have to be banished from our thinking.

---

[78] a Merriam-Webster definition.
[79] a Merriam-Webster number 2 definition.

Joe: but herb, did not the (re)discovered science of quantum mechanics necessarily suppose some non-physical realms? How could the scientists who used quantum mechanics be materialists?

herb: thank you my dear Joe. Indeed, while quantum mechanical theories do underly chemistry / biology / and life forms in general (and electromagnetics should also be included, i.e., the science behind the operations of the mysterious fluid type magical substance called electricity), it is true that many scientists that utilize quantum mechanical techniques do not consider the fact that quantum mechanics requires a non-physical realm.

One very important aspect of quantum mechanics is the so-called *quantum vacuum*. This is the lowest level from whence / where (is it a place?) things of the / this physical Universe need interact with 'things' that pop in-and-out of the physical Universe.

Joe: and that requires, as we have discussed before, a definite need for a non-physical realm.

AlBe: so the non-physical realm is where (is it a place?) the 'things' that pop in-and-out of the physical Universe are located (damn, it is probably not a place?) when not in physical Universe?

herb: yes, indeed, my dears Joe and AlBe; butt you see there are some subtleties involved. For you my dear AlBe, try to think about the possibility of a non-spatio-temporal abode (an abode where our normal ideas about spatial extent, or the passage of time, may not have any meaning).

Sophia: what else occupies this non-spatio-temporal abode(s) my dear herb? Is it where (doh, it is not a place!) the wild Things Archons roam?

herb: I would contend my dear (the goddess) Sophia, that this is most surely the case. Also, however, all is not to be feared (nor in point of fact, is anything to be feared... no ego, no envy, no fear) as it concerns the non-physical, as this is also the abode that most likely facilitates the actions associated with the immortality of the soul.

thalamus nudges herb and suggests they should save some wood for tomorrow - if there is to be a tomorrow, and maybe continue the conversation in the morning

the interlocutors each access their black boxes - that glorious idea of the great shaman Bucky - and commence to construct their sleeping and shelter arrangements; all the while a soft background music is being strum by the goddess

an Objective reality?

Glia, Joe, and herb have been busy preparing a morning meal, which includes eggs with green olives, over a fire that thalamus has urged back into life for this exact purpose. After the meal, a morning hike, and various waste processing needs are attended to, the discussion continues

Giordano: my dear herb, we were last talking about quantum mechanics and the immortality of the soul; and do you think this is a key point in understanding 'What is'?

Glia: pardon my interruption my dears Giordano and herb, butt were we not also talking about the conspiracy of 'materialism'?

herb: quite right, both of you; for the conspiracy of 'materialism' is very germane to understanding 'What is.'

I would contend that one of the biggest questions an individual needs ask themselves before they even bother to start tackling higher spiritual matters, is the question as to whether the / this physical Universe is a 'real' thing at all, or if in fact, it is only a dream of sorts. To wit; there is many a philosophy that would suppose that the / this Universe that an individual finds themselves in is just a 'subjective' experience (for that individual), and that the / this Universe is more of a masquerade than anything else.

Joe: my dear herb, was this idea of the / this Universe being nothing more than a type of masquerade one of the options presented in the classical metaphysics of the *Devil's World*, where every thing / thought / feeling was feed to the individual as if they were just a brain-in-a-vat?

herb: quite right my dear Joe, and this is why we should take a stance in our viewpoints, and ask the question as to whether we can define / delineate an Objective reality[80]

Note, as we move forward in our discussions, we will happen upon a very related concept, wherein many moderns view the human experience as *epiphemeral* – meaning that our experience is like a old-time motion picture show, and that it only appears that we are interacting in a cause-and-effect manner.

Sophia: my dear herb, if the placement of humans (and you k-now who you are) is in an Objective reality as opposed to your previously mentioned, more subjective 'thing-y,' then will this Objective reality contain both a physical and non-physical?

herb: yes my dear (the goddess) Sophia, both are required, and this is where quantum mechanics will help us. The shaman Wheeler exclaimed after much thought and reflection

> I found myself forced to invent the idea of quantum foam, made up not merely of particles popping into and out of existence without limit, but of space-time itself churned into a lather of distorted geometry.

Thus, when contemplating the possibility of an Objective reality, I submit it may be best if the / this physical Universe is imagined as a seething dance of energy events whose true dynamics should be couched in terms of Planck lengths, Planck time[81], and Wheeler's quantum foam, and whose quantum description would include so-called virtual particles (and anti-matter) which seemingly pass in and out of physical existence and that may even deploy effects both Forward and backward in time.

---

[80] even though, most interlocutors would have to agree, it is quite absurd to think that humans (and you k-now who you are) could ever really k-now reality... butt on the other-cheek, the great shaman RAW did remind us that 'reality' is a word

[81] The Planck scale is the proposed bottom-level for physical events.

both Forward and backward in time

> *thalamus has been walking among the interlocutors offering more coffee - more coffee - and on cue, a few of the interlocutors, including herb, recuse themselves for some moments and take a smoke break. The discussion, however, continues*

thalamus: eye have heard herb (hhh) discuss the idea and calculations of the quantum mechanics quite often, and the gist of the idea is that this is the peculiar part of modern physical theory where we / the scientists should come to terms with the needed involvement of negative energy electrons, and other antiparticles.

Sophia: is this where the shaman Dirac's idea of a(n) (inexhaustible) 'sea of negative energy electrons' comes into play?

thalamus: yes my dear (the goddess) Sophia, Dirac's 'sea,' and many other ideas come into play. After the first wave of quantum theories, others continued to postulate further ideas, including the famous diagrams of the shaman Feynman in his QED[82] theories, where we find that all the virtual photons and antiparticles are (again) assumed to literally jump into and out of existence (within, of course, time and energy parameters given by the shaman Heisenberg 's *uncertainty principle*).

Sophia: is this what herb speaks of when he uses the term 'reality flux'?

Joe: yes, and herb is insistent that the *reality flux* is in fact required in order for (physical) process to exist in the / this physical Universe.

thalamus: quite right my dears Joe and (the goddess) Sophia. And here was the funny - or not so funny - part of it; the quantum scientists knew these basics also, but they did not want to highlight the idea of 'reality flux'.

Giordano: why? Because they did not want to have to deal

---

[82] QED is short for quantum electrodynamics, the physics of electron / photon interaction.

with the important / next / obvious question of where / whence does the stuff of the 'reality flux' go to and come from (doh, it is not a place)?

thalamus: indeed my dear Giordano, it is the obvious question. Butt what is a materialist to do? (Go against the wishes of the sons of Belial (those sOb's) and talk / write about spiritual matters that matter?)

Joe: quite the conundrum.

a Jitterbug of space

thalamus: herb uses the term *Negative Universe* to describe that aspect[83] of the quantum vacuum where antiparticles flux to-and-fro when they (i.e., the antiparticles) are not in the physical universe.

Sophia: why call the abode the 'Negative Universe,' and not the *noumenal*, or the astral / ethereal, as has been done by many others?

thalamus: yes, my dear (the goddess) Sophia, this was an option for herb when, during his studies on how the Universe computes, he happened upon the absolute requirement that there be such a realm.

It was, I suppose, in homage to the great shaman R. Buckminster Fuller (Bucky) that herb has decided to adopt the name Negative Universe for this required abode.

Joe: right, I have heard herb explain his rationale for this. He tells a story about how the great shaman Bucky used Negative Universe for the realm where the higher order / frequency geometrical forms reside, and Bucky also postulated that this is where (is it a place?) thought forms reside. When herb put the two together (i.e., the quantum vacuum required non-spatio-temporal realm, and an abode wherein 'Mind' could reside), it was most appropriate to use the term Negative Universe.

thalamus: and his hope was that it would not seem so new age'y, and that it could possibly entice scholars / researchers / readers to investigate on their own 'The World of Buckminster Fuller'[84].

---

[83] calling 'Negative Universe' a 'place' would imply a spatial essence
[84] which is a great youtube video!

Sophia: so then, 'reality flux' is the part of the quantum vacuum that connects Negative Universe to the / this physical Universe?
thalamus: yes.
Sophia: and this is where herb gets the idea that 'reality flux' is the power source for Universe (as computation)?
thalamus and Joe: yes, exactly.
Giordano: so if we back up then to the shaman Wheeler's quantum foam, is that in the / this physical Universe, or is the foam part of the 'reality flux'?
thalamus: very nice my dear Giordano, for your keen interest and acute attention to these matters, has brought us to the point where we can postulate that the quantum foam is possibly the interface / membrane between 'reality flux' and the /this physical Universe.

Also, it is possible that the foam is more related to the ancient idea of the *Pleroma*, which for a lack of a better description, is that substance (is it a substance?) that is formed, to present, to those who would observe, the / this physical Universe.

Joe: right, butt we should not forget the geometry of space aspect to all this my dear thalamus. As herb has conjectured, it is the geometrical aspect[85] that 'tells'[86] the Pleroma how to form, and that is what led herb to ponder - often - that the foam level (of the / this physical realm?) is the fabled Pleroma.

thalamus: of course, and many thanks my dear Joe. Because at this membrane level, herb has indeed insisted we consider the fact that space (in the / this physical Universe) has a geometry (as was supposed by the great shaman R. Buckinster Fuller - and others, including the great shaman Plato). Further, as you so well intimated my dear Joe, it is the synergetic geometry of the great shaman R. Buckminster Fuller that can be employed here to provide a scaffolding of sorts for both the 'things' of the / this physical Universe, and for the organization of the 'space' itself. The concept of tensegrity is also important here, and elsewhere.

Joe: and the jitterbug, do not forget the jitterbug!

---

[85] so beautifully analyzed by the great shaman Bucky.
[86] as an ancient Aeon is often responsible to do.

Sophia: the jitterbug? is that the great shaman Bucky's term for when the VE (vector equilibrium / cubeoctahedron) changes to an icosahedron, and then back again to a VE... in that beautiful, geometrical (somewhat harmonic) motion[87]?

thalamus: yes, that jitterbug my dear (the goddess) Sophia. herb has speculated that Wheeler's bubbling (bubbling bubbling - bbb) foam is in fact 'built'-with micro-VE[88], and that the creation of virtual VE (via jitterbug actions of the Pleroma - powered by 'reality flux') in response / cooperation, as the 'things' in 'space' evolve, is the mechanism for space creation[89] - thus space 'expands' from witheverywhere.[90]

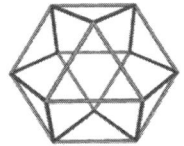  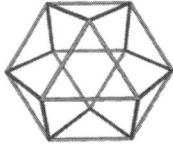 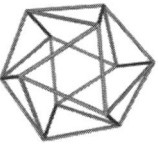

the Secret teaching of all ages

herb, Glia, and AlBe have returned from their break, and thalamus has quickly updated herb on the discourse that he had missed

herb: I would like to make sure that my dear Giordano gets this other key point that belongs here; and that provides more evidence towards understanding Negative Universe as a fundamental part of Universe.

My dear Giordano, you are familiar with the investigations into the science of consciousness by the shaman Hammeroff and the shaman Penrose?

---

[87] as shown in the graphic - courtesy the shaman RW Gray
[88] VE reads as vector equilibrium, that 12-around-1 arrangement which incorporates 4 inter-penetrating hexagonal planes.
[89] as opposed to all that silliness about dark energy and dark matter.
[90] this jitterbug-ing of the pleroma and the associated virtual VE creation could very well explain a mechanism for momentum conservation - as herb has suggested; thus, in a way, saying that 'things' both 'carry' their 'space' with them, while all the time causing (activating?) new 'space' creation due to a Divine interaction with the pleroma.

Giordano: you k-now I am my dear herb, as we have both discussed those investigations wherein they place the microtublules (within the neurons) at the intersection between their Orch OR quantum collapse and brain functioning - their explanation for how mental process intervenes upon the physical world[91].

herb: precisely; then think of Negative Universe as the same type of abode as the shaman Penrose's 'Platonic World,' which as such is envisioned as a repository for the extra stuff (beyond the physical) that is needed for mental activity[92].

Sophia: right, butt how do we then explain that the shaman Penrose remained a bit of a materialist, in that even though he intuited the reality of the situation, in that 'Mind' required more than just what we have in the physical, he did not / would not take the obvious leap to consider the Secret teaching of all ages which has always pointed to the associated reality of a non-physical.

Giordano: maybe he was just too respectable to be seen as a Cartesian dualist, where, for example, as the shaman Jaegwon Kim points out (that for the dualist)

> the world is split in two with Minds on one side and stuff on the other.

Evidently, the shaman Penrose simply could not shake his materialist roots.

> *the interlocutors laugh a bit at the realization that no matter how one tries to think clearly about things, we are still bound in ways that sometimes, we can only suppose, only the Divine can understand*

herb: eye brought up the topic of the great shaman Manly P Hall's *The Secret Teaching of all Ages*, and I would like to say

---

[91] which as rightly noted by others would imply that other entities that employ microtubules in their construction - as it is with many plants - can also enjoy / be burdened with consciousness.

[92] recall here that the shaman Penrose famously tried to 'prove' to his contemporaries that mental activity could not be solely the result of only physical process; thus the need for something extra - whence appears the use of his so-called 'Platonic World' (which is beyond?, or embedded deep within, the physical – in the form of convenient caches in the space-time geometry – Wheeler's foam?).

more about the subtleties of the organization of the non-physical realms that were surely in some times taken as a given, and in other times taken only by 'fools.'

Sophia: surely my dear herb, you must include the idea and mechanism(s) of reincarnation into these aforementioned subtleties associated with the non-physical realms (as these realms are surely supportive of the 'passage' and 'movements' (is it a place?) of each immortal soul).

herb: and why would that be my dear (the goddess) Sophia, is that an idea that that was surely in some times taken as a given, and in other times taken only by 'fools'?

Sophia: you k-now only too well my dear herb that this was indeed a fundamental tenet of many religious systems / practices / reality-tunnels[93] of past times. Including systems enjoyed by the Egyptians, the Zoroasters, and Jesus even (as evidenced in the Gnostic gospel of Thomas).

Giordano: was this idea of reincarnation expunged from modern times due to a lack of teaching?

herb: my my my (mmm), my dear Giordano; is that what you think?

Giordano: forgive me my dear herb, as even I can plainly see that any discussion of this issue (concerning reincarnation) in our time must reckon both with the Gnostic reasonings, and the machinations of the sons of Belial (those sOb's).

AlBe: my dear herb, if everyone is forced into reincarnation (time after time, after time, after time, ...), then this surely is the doings of some Gnostic demiurge(s); for when could any (immortal?) soul find refuge from the / this physical world? It just seems so unreasonable, since one would have to come back over and over, with a memory wipe in between, and while I do not want to suppose to question the 'intentions' of the Divine, it simply does not seem fair.

herb: thank you both my dears AlBe and Giordano for bringing forth these very important points. Now, consider this closely ...

---

[93] a RAW and Leary philosophical construction - and an important one at that.

on Some theoretical problems with brain emulation

> *thalamus had interrupted herb before he started into, what he and Joe called the Stockholm considerations, and asked him to consider a meal break before they continued the discussion*
>
> *After the meal and potty breaks, the interlocutors began a hike further down the hidden forest path into what many legends thought to be an enchanted forest (realm?)*

herb: (lighting a cigarette as they walk the path) the Egyptians, in some of the history that still survives (even if extremely distorted), had many rituals dedicated to overcoming the reincarnation cycle, so that one may move past the Aeons / Archons of the astral realms and return to the source whence our immortal souls once resided, the realm(s) of the transcendent Divine. These enchantments are still available to us in the divine magical tradition of theurgy.

The Gnostics wrote at length about the trappings of the / this physical realm, and used the allegory of sleep, or drunkenness, to explain our contentment with the / this physical world. This may happen even though one may well k-now of the true history / science / structure of the / this physical realm. Then, they could perhaps be numb enough to be fooled when they enter the non-physical realm(s), and consent to a request that they are required to return to the / this physical realm. Typically this return is accompanied by a 'sort of' 'mind-wipe,' but there are enchantments related to that also.

You see, here is where the sons of Belial (those sOb's) have so many recent triumphs, in that through propaganda and control manuals they have 'convinced' / 'shamed' / bedeviled these incarnated immortal souls to believe in

- karma – something designed to convince one to return,
- signed pacts – with spirits / Aeons / Archons / demiurge(s) that evidently might be said to have been agreed in a realm for which the soul, may in fact, have no 'memory' of 'signing', or
- a judgment hearing – that can be simulated for the recently departed (may the Divine bless the / this soul) as the sOb's well k-now.

The use of any of these examples increases the likelihood that the immortal soul will submit to the request / demand that they once again become incarnated into the physical realm, via what has oft been described as reincarnation.

Sophia: my dear herb, you may exaggerate a bit, even though I k-now you mean well.

herb: my dear (the goddess) Sophia, you, among all sentient entities throughout (is it a space?) the physical and non-physical, k-now well of the demiurge(s) methods, and the results we have witnessed due to unceasing propaganda of materialism, and the denial of the reality of the non-physical and the denial / forgetting as it concerns the immortality of the soul.

Sophia: granted my dear herb, butt is not there still more? What if, for example, souls en masse 'decided' not to return to the physical realm (via reincarnation)?

herb: very nice my dear (the goddess) Sophia. So we have basically two issues I would like to clarify on these recent topics (dealing with the possible processes concerned with the how of why immortal souls can be entrapped in the / this physical realm).

Consider again the issue of reincarnation, and the suggestion that if an immortal soul is *always* required to return, then the (public) Gnostic teachings got it spot on when they label the / this physical realm a prison of sorts. However, consider, the stance / religious belief / dogma offered up on this particular dilemma by a very large religion on the planet; that is, the Hindu reincarnation process requires some 480+ incarnations into the physical realm before the soul is no longer required to return. In this way, we could view the / this physical Universe (built, mind you, by a demiurge(s)), as a type of training ground.

Sophia: during this long cycle my dear herb, do the soul(s) decide into which physical incarnation they will go?

herb: eye k-now not my dear (the goddess) Sophia whether a child can help where they are born. I have often pondered the question however, and at one time this idea (that a child could not help where they were born) was a candidate for a 'Universal Truth'[94].

Now, the second issue I wanted to discuss relates to the quest that occupied many a scientist in our most recent past, where it was the Brain builders dream to bring forth consciousness into a machine / substrate different from a human brain.

Giordano: indeed, these brain-builders employed techniques, some related to artificial intelligence, other related to neural nets and deep learning, in efforts to create sentient / thinking machines.

thalamus: and some of them employed brain connection mapping with the most elaborate of machines. You see, these machines would thinly slice away at brain matter, and basically try to unravel the synaptic connections all the while keeping track of these connections in very elaborate programs.

herb: indeed, and as some of you k-now, it has always been my contention, that we k-new the reason why when they turned on their brain machines in those deep / dark / double top sekret labs, the brain machine was always psychotic when they turned it on.

Joe: right, my dear herb, because 'Mind' and brain are in two different realms, and a brain – whether machine substrated, or biological substrated, in order to exhibit true consciousness, must be connected to a 'Mind' in Negative Universe.

herb: this indeed has been our contention my dear Joe. And dig this, if the sons of Belial (those sOb's) had been successful in entrapping an immortal soul from (is it a place?) Negative Universe, then they would be mimicking their buddy, the Gnostic demiurge(s), in that they would be entrapping 'Mind' / soul in something of their own creation.

---

[94] a Universal Truth is one that admits no non-zero derivatives.

wild Grey sweet earl orange

all the interlocutors sit for a while, a few with their mouth agape, as the band begins to belt out, in a folksy funk type way, the classic 'Rock and Roll,' which is appreciated by all. Later in the evening by a makeshift campfire...

Giordano: my dear herb, while I can usually follow the reasonings that bring us to some conclusions here and there, it seems so much of the claims here and there about reincarnation, the immortality of the soul, et. al., can be nothing more than the opinion of a man.

Sophia: i must concur with our dear Giordano here my dear herb, and ask: what evidence can you present to back your story telling as it concerns the immortality of the soul?

herb: thank you for your honest and probing analysis my dear Giordano and my dear (the goddess) Sophia. To explain my stance on these topics we should (re)visit some ideas I will attribute to the great shaman Robert Anton Wilson (RAW).

thalamus: RAW wrote those many books, mostly satire(?), that made fun of / probed the 'Illuminati' mythos?

herb: much more than that my dear thalamus. RAW, the self-proclaimed guerrilla ontologist, who was a great fan of the great shaman Bucky, also railed against science taking itself too seriously, thus he was an opponent of 'irrational materialism.'

RAW would often use the term 'reality-tunnel' that he and the shaman Leary came to use to describe the attitude / experience filters that an individual builds up and uses over time.

Sophia: if I may dear herb?

herb: please, (the goddess) Sophia, break it down for us

Sophia: the great shaman RAW was able to step back from the, what easily could be, an overwhelming reality-feed quite well, and, as the great shaman Bucky did, take a stance that the / this physical Universe is a conglomeration of non-simultaneous overlapping of energy events. In this way, they had tackled the same ontological question as you my dear herb when you posit the query: 'an Objective reality?'.

the great shaman RAW also studied the quantum mechanics and the dream / astral realms[95], and k-new that many ancient traditions held the same beliefs that are currently found in the

---

[95] RAW allowed towards the end, a 'copy what you want' attitude, thus bear witness, to one of his enchanting tales: – FNORD 'Very nice,' I said. 'But why did you bring me up here?' 'It's time for you to see the fnords,' he replied. Then I woke up in bed and it was the next morning. I made breakfast in a pretty nasty mood, wondering if I'd seen the fnords, whatever the hell they were, in the hours he had blacked out, or if I would see them as soon as I went out into the street. I had some pretty gruesome ideas about them, I must admit. Creatures with three eyes and tentacles, survivors from Atlantis, who walked among us, invisible due to some form of mind shield, and did hideous work for the Illuminati. It was unnerving to contemplate, and I finally gave in to my fears and peeked out the window, thinking it might be better to see them from a distance first. Nothing. Just ordinary sleepy people, heading for their busses and subways. That calmed me a little, so I set out the toast and coffee and fetched the New York Times from the hallway. I turned the radio to WBAI and caught some good Vivaldi, sat down, grabbed a piece of toast and started skimming the first page. Then I saw the fnords. The feature story involved another of the endless squabbles between Russia and the U.S. in the UN General Assembly, and after each direct quote from the Russian delegate I read a quite distinct "Fnord!" The second lead was about a debate in congress on getting the troops out of Costa Rica; every argument presented by Senator Bacon was followed by another "Fnord!" At the bottom of the page was a Times depth-type study of the growing pollution problem and the increasing use of gas masks among New Yorkers; the most distressing chemical facts were interpolated with more "Fnords." Suddenly I saw Hagbard's eyes burning into me and heard his voice: "Your heart will remain calm. Your adrenalin gland will remain calm. Calm, all-over calm. You will not panic. you will look at the fnord and see the it. You will not evade it or black it out. you will stay calm and face it." And further back, way back: my first-grade teacher writing FNORD on the blackboard, while a wheel with a spiral design turned and turned on his desk, turned and turned, and his voice droned on, IF YOU DON'T SEE THE FNORD IT CAN'T EAT YOU, DON'T SEE THE FNORD, DON'T SEE THE FNORD . . . I looked back at the paper and still saw the fnords. This was one step beyond Pavlov, I realized. The first conditioned reflex was to experience the panic reaction (the activation syndrome, it's technically called) whenever encountering the word "fnord." The second conditioned reflex was to black out what happened, including the word itself, and just to feel a general low-grade emergency without knowing why. And the third step, of course, was to attribute this anxiety to the news stories, which were bad enough in themselves anyway. Of course, the essence of control is fear. The fnords produced a whole population walking around in chronic low-grade emergency, tormented by ulcers, dizzy spells, nightmares, heart palpitations and all the other symptoms of too much adrenalin. All my left-wing arrogance and contempt for my countrymen melted, and I felt a genuine pity. No wonder the poor bastards believe anything they're told, walk through pollution and overcrowding without complaining, watch their son hauled off to endless wars and butchered, never protest, never fight back, never show much happiness or eroticism or curiosity or normal human emotion, live with perpetual tunnel vision, walk past a slum without seeing either the human misery it contains or the potential threat it poses to their security . . . Then I got a hunch, and turned quickly to the advertisements. it was as I expected: no fnords. That was part of the gimmick, too: only in consumption, endless consumption, could they escape the amorphous threat of the invisible fnords. I kept thinking about it on my way to the office. If I pointed out a fnord to somebody who hadn't been deconditioned, as Hagbard deconditioned me, what would he or she say? They'd probably read the word before or after it. "No this word," I'd say. And they would again read an adjacent word. But would their panic level rise as the threat came closer to consciousness? I preferred not to try the experiment; it might have ended with a psychotic fugue in the subject. The conditioning, after all, went back to grade school. No wonder we all hate those teachers so much: we have a dim, masked memory of what they've done to us in converting us into good and faithful servants for the Illuminati.

consequences of quantum mechanical theories (as you herb have discussed, for example, when we find in quantum mechanics the obvious need for a non-physical). From some recent discussions, concerning the influence / involvement of the 'reality flux,' the great shaman RAW understood interactions were taking place on / at many levels within the modern human (and you k-now who you are).

thalamus: so not only is our 'Mind' interacting with Universe, but our physical bodies also, right?

Sophia: indeed my dear thalamus. So here then is the issue: if we tried to be 'aware' of all our interactions with the / this physical Universe and Negative Universe, the multitude of signals would completely overload the individual, and all would seem extremely chaotic (as the true 'reality' of all things must surely seem to be).

Thus, the great shaman RAW and the shaman Leary came to speak of a 'reality-tunnel' that each individual must in fact build in order to function at all in the / this physical Universe. The individual may over time become more aware of all / some of these inputs (and it helps if the individual is looking for some – all the better chance to detect the some).

herb: so this is why I wanted (the goddess) Sophia to explain this analysis of the great shaman RAW, as it helps explain how I could come to have these stances concerning, for example, reincarnation and the immortality of the soul... it is information that I have gathered because I was looking for it.

My education and experience built up a 'reality-tunnel' that had me pay attention to the geometrical and energetic / information needs for a physical Universe in which 'Mind' is supported. I have adopted the legend of the immortality of the soul from *The Secret Teaching of All Ages* because that fits into a part of my 'reality-tunnel' that requires / demands that the Universe make some sense; plus, by the way, it helps with the question of 'why bad things happen to good people.'

Giordano: so that your claims, while - of course - the opinion of man, are supported by the calibration of your individual 'reality-tunnel' as it concerns these most basic of questions / concerns.

herb: and when eye say that the Universe needs to make sense, this addresses one of the issues the great shaman Pascal dealt with when thinking man 'lost in the physical Universe, a place indifferent to humans' (and you k-now who you are); and eye k-now (k)not if the great shaman Pascal recanted, or continued in this nihilist path; but I picked another reality-tunnel where in fact eye expect some meaning may (or June) be found

Sophia: my dear Giordano, these conclusions / stances that herb takes are indeed the opinion of man, but are also to be found - as herb intimates - throughout all ancient times, and in other civilizations throughout the / this physical Universe.

herb: and, if you would pardon me, my dear (the goddess) Sophia, not only are many of these ideas, that we have discussed, old ideas from different languages and different times, but the use of modern mathematics and science also point, many times implicitly, to the fact that sometimes old ideas are the best ideas (or at least remembered ideas - surely to the disappointment of the sons of Belial (those sOb's)).

Butt under no circumstances, should you ever think that these ideas fully describe the mystery that exists right in front of your face (and anyone else's face who would dare to not have eyes Wide shut).

Sophia: another key area of the great shaman RAW's thinking that is germane to our original query of describing 'What is', is his straight up uncertainty concerning the Objective reality of spirits / Aeons ? Archons (and their ilk). You see, the great shaman RAW had adopted the great shaman Israel Regardie's suggestion that we not try too hard to decide whether all the spiritual influences (both Divine and dark) are simply part of an individuals psyche, or are some 'thing' separately objective... they (RAW and Regardie) thus decided to leave that question open (so as not to get lost on that trail one can only surmise... or out of respect for the mystery... or...).

herb: and I always (seemingly) adopt the objective 'reality' stance whenever given the chance, and move forward with my 'reality-tunnel' programming from there.

it Is a great mystery

> *the interlocutors, after hiking for who k-nows how long, find themselves in what seems like a previously undiscovered clearing, deep - they k-now not how far - into the enchanted forest. They notice the ruins slightly hidden here and there, and wonder if something still lives near*

herb: the next thing we should consider as we continue again investigating the 'conspiracy of materialism' is the concept / mechanisms / beauty associated with 'life.' And by 'life' I would like to engender ideas associated with the Platonic flame that 'breathes' an essence into all 'things.'

Sophia: so you are attempting here my dear herb to talk about all 'life': the plants, the animals, the humans (and they k-now who they are)... anything where one could say that matter is animated in some way, and thus can experience growth and decay, reproduction and death?

herb: indeed my dear (the goddess) Sophia, this is the next topic I would like us all to peruse, as it does involve the interaction of the physical and the non-physical realms (and is thus still within the purview of the 'conspiracy of materialism' - the second 1/2 of our Pythagorean basics).

AlBe: will we be discussing herewith the 'miracle of birth'? As this was always, and continues to be, one candidate for what you would consider the most mysterious thing of all!

herb: yes my dear AlBe, butt I think maybe we / you have just done what you queried about. To elaborate, the 'miracle of birth,' which Is a great mystery, in the animal kingdom is miraculous enough (as it is in the plant kingdom, which brings forth the teleological mystery of the magical seed); however it is absolutely unbelievable when we contemplate a human birth, where a new individual (from an old immortal soul) is brought forth into the / this physical Universe. And even though from a Gnostic conception, it may be interpreted as a mistake to entrap an immortal soul in the / this physical Universe; the mystery of it all remains.

> *thalamus offers the suggestion that they should gather wood and berries, and camp here for the night - if not a few days*

that Life thing

> *it has been a few days since the interlocutors had enjoyed deep conversation as it concerns 'What is.' The ruins newly found show / indicate ancient signs of thinking about patterns and integrities – music to the interlocutors soul... a soft sound that they all wish everyone could hear*

herb: this 'life' thing requires both the physical substrate of molecules that make up a body, and a 'reality flux' (which is part of the quantum vacuum) to keep the system running. As you may recall, as mentioned in discussions long ago, one can be sure that any interactions involving electrons needs 'stuff' popping into-and-out of existence in order for the interactions to occur.

We should also take into account various possibilities of organizing principles mediated by the 'reality flux,' which would include morphic resonance, espoused by the shaman Sheldrake, and the Askashic records mythos that supposes a somewhat time-independent access to goings on's in the physical realm.

Negative Universe is also an appropriate abode wherein 'Mind' can reside. Then 'Mind' interacts with the brain in physical Universe through mechanisms of the 'reality flux' (as we have discussed before), which is similar to the shaman Hammeroff and the shaman Penrose's ideas when it is the quantum interactions in microtubule within the brain that bring forth conscious experience.

Glia: what about our dream worlds? Are they part of the goings on in the Negative Universe also my dear herb?

herb: most likely, and here is where we should highlight again the fact that the Negative Universe may have structure.

Sophia: the Secret teaching of all ages has always included the shamanic notions of structure associated with the astral realms beyond the physical realms, which much like the ideas of the Sephiroth (brought to us from the sons of Beliel - those sOb's), are abodes which under current considerations would fit nicely into what you are designating as Negative Universe my dear herb.

thalamus: what would a materialist scientist have us believe is the source of our dream worlds, or the possibility of other realms for that matter?

Glia: i would suppose they would claim the dream as nothing but chatter / noise within the physical nervous system that is connected to ones brain.

Giordano: and very likely call you insane for supposing that there is anything beyond the physical... in contradistinction to what some believe to have been a more natural religion, enjoyed, for example by some ancient Egyptians, as (the goddess) Sophia explained in my book "The Expulsion of the Triumphant Beast[96]" we could rather have respect for the sane wo/man

> `... *who contains within himself a great doctrine and judgement of natural and magic things, concerning the various reasons through which form and divine substance either immerse or enfold, or distribute themselves through all, with all, and from all ...*'

herb: so that Life thing, such subtle manipulations it seems of that web of K-nots of 'light,' requires Negative Universe to give it direction, along the course of a possible 12 ways

---

[96] a book by our dear Giordano Bruno

conspiracy Of hidden history

(the goddess) Sophia has agreed to herb's request to give some insight into how little credence one can assign to written historical accounts

Sophia: in recent historical times, the shaman Dewey Larson noted that

> Somewhere along the line, that which is true is being made to appear false, because that which is false is accepted as truth.
> - Dewey Larson

and sadly, this was not a new situation, because, as you see, the sons of Beliel (those sOb's) have always twisted written accounts in a ceaseless propaganda campaign against humans (and you k-now who you are).

Giordano: can you give us some examples (the goddess) Sophia?

Sophia: a funny example, or not, depending on how one views genocide I suppose, is an example involving the great shaman Simon Magus. The scene is a battle between the great shaman and (the liar) Peter. As told in the ancient control manual, it was (the liar) Peter who while engaged in a magical battle with the great shaman Simon Magus, brought the great shaman down from his flight so that he splattered on the ground causing his death.

We k-now now (nono) that this foretold / celebrated the sOb's attempted genocide of the Gnostics during those times when that particular control manual was being put together.

The many stories of Druids / Celts / Merlin and associated ilk who moved great stones and built great monuments is similarly shrouded with propaganda by the sOb's, who continue, most likely to this day, to eradicate all records of anything that does not agree with the false narratives that the sOb's have been presenting to humans (and you k-now who you are) for far too long.

herb: can you tell us, (the goddess) Sophia, what you k-now, or are allowed to tell us about the sorcerer John Dee?

Sophia: the sorcerer John Dee was a dark magus (and there were many) who assisted one of the war mongering queens in the recent past of humans (and you k-now who you are) on this planet.

herb: did not this dark sorcerer play a role in the construction of the first official English language, the construction of a version of one of the key control manuals (the KJV), and last, but not least, the reintroduction of the wicked sorcery k-nown as Enochian magic / but witch had many other names, many other variants, at many other times.

Joe: does not the Enochian magic of the sorcerer John Dee go back to the great shaman Soloman, who, as you have intimated in the past my dear herb (and which has been suggested by others), attempted to use related techniques of the Enochian magic to entrap the Archons for all time, and thus forbid the interactions of Archons and humans (and you k-now who you are) forever more

Fate: if I may dear herb and my dear Joe, and with reverence to (the goddess) Sophia, who may be a bit tongue tied in this situation, I would like to comment on the hidden history of the dark sorcerer John Dee

Giordano: when one says 'dark', what does that imply? One k-nows, as we are reminded quite often, that the control manuals forbade any magical practices; but does this mean that all of them (i.e., magical practices) are 'dark'?

Fate: my dear Giordano, have you not discussed in the past how magick is cleaved into two (2), almost disjoint, categories (or branches - but not of trees)? One category can be labeled as sorcery, and the other (category / branch - but not of a tree) is in recent times given the divine label theurgy.

Sophia: and herb has indeed, as you query, my dear (the goddess) Fate, given background on the distinction between theurgy and sorcery.

a Dark magus of history

AlBe, while entranced somewhat by the openness and story of the battle between sorcerers, their destiny, all of course in association with (the goddess) Necessity, insists to the interlocutors that a break is in order. With an appropriate three breaths and the associated stretching at the end, discussion of the Dark magus, and his many machinations, commences again

herb: while the sorcerer John Dee is not the most malevolent dark magus of history, his case in our false history is the / a classic 'in your face' study of how humans (and you k-now who you are) are manipulated.

We must, in order to grasp this idea, for example, understand that one is more likely than not to be ridiculed when connecting the lineage of the sorcerer John Dee with the sons of Beliel (those sOb's). There has been, before the sorcerer John Dee's time, during, and afterwards, a / the constant struggle / propaganda of the sOb's to keep down the 'alien man'[97]

Fate: the sorcerer John Dee did in fact have access to the ancient records and techniques of the sOb's. In his times, along with the shaman (sometimes sorcerer, and sometimes theurgist) Francis Bacon, introduced many a technique that had long been in the works.

Glia: they, these two sorcerers, Dee and Bacon, brought forth many techniques in mind control. Along with their English language, they forced upon us, with spells (their words), a prototype for what the sons of Beliel (those sOb's) desired human nature to become. For example, just consider for a moment those hideous works, because of their normative effects on the most base thoughts that ever occur within wo/man, those works attributed to Shakespeare (a hapless demi-muse for the sorcerer Sir Francis Bacon), have led the way to a (programmed) belief that this is how humans are supposed to behave

---

[96] A Gnostic term for those sons of the Serpent who attempted, against all odds, to save the newer slaves of (the turned dark) Atlantis in those very trying times of old - before the water goddess Tiamat lost the cosmic battle, which stripped from her the deep waters that eventually fell to the third stone from the sun

Fate: as Fate would have it, and a timeless sorrow and apology to you, and all, my dear Glia; but it is not only destiny, but also necessity that brings forth activities into this / the physical realm.

All these combined 'words' that the sOb's introduced / refined at those times could indeed put shackles (if you will) on the 'Mind' activities and consciousness in general, if it nudged a human (and you k-now who you are) to forget that consciousness and thought can be cast in other ways, using different 'terms' than just words.

thalamus: and these words in the English can work other types of sorcery also, can it not?

Fate: well done my dear thalamus, as indeed, the sophistication of their spell system (the English language and the great memes foisted onto the minds of wo/men by the sOb's) is all the more impressive when the same spoken word can have multiple spelling sequences and multiple meanings.

herb: eye write witch means that eye m right.

Fate: the power of suggestion, to which many are susceptible, and a written (false) history is a bane to the hope of divine congress.

herb: butt, it is achievable, is it not my dear (the goddess) Fate?

Fate: of course my dear herb, and as you k-now well, the shaman RAW often insisted that 'you can change the way you are thinking right where you are standing now.'

herb: indeed. And what other accomplishments of the Dark magus John Dee should we enumerate before moving forward (or backward)?

Sophia: the Dark magus John Dee, through techniques of his Enochian magic, made somewhat infamous by both Marlowe and Goethe with their narratives of the (sometimes) good Doctor Faust(us), was able to get what would be the key instructions for the sOb's for their next 500 years of work; because as you see, the Dark magus was informed that the sOb's needed to master the use of the crystal silicon, and gain the techniques to put together matter in such a manner that 'life' can be breathed into it.

herb: babylonian Golem magick then

a Wandering who?

> *herb, after a private conversation with one of the many statues among the ruins, as is an ancient tradition among theurgists, brings more hidden history to the fore, and he selects a topic often discussed by the shaman Gilad Atzmon - one of their tribe mind you*

herb: the story, written records, in an older tongue than English, of the phenomena of Judaism, also has a (not so) hidden history, does it not my dear (the goddess) Fate?

Sophia: if I may my dear (the goddess) Fate, add that the magical Hebrew language can cast different types of thoughts / fetishes / reality (not so real) tunnels on the unk-nowing human (and you k-now who you are) than can the spell friendly, sometimes meaningless (by design) English language.

Fate: and as Fate would have it, and apologies again, these ancient books of some of the wandering sons of Beliel (those sOb's) were forgeries and plagiarism of the highest order. Taking liberties with Babylonian, Chaldren, Egyptian, and Alexandrian mythos, the sOb's cast a story and (fake) backdrop for their history.

herb: here then, for example, the Torah co-opted / recounted in a dumb-downed fashion the legend of the shaman Giglimesh. Worse still, they reversed the role of the Hyskos tribe in taking over the politics of Egypt, and then their subsequent expulsion from those lands; for you see, Egypt did not k-now slavery until the Hyskos brought that debauchery with them.

Sophia: even though free masonic tradition can retrace these steps, and the battle between the demi-sOb's and the true thinkers in those ancient lands, it hides much of the story in their lore - mostly undecipherable, it is also without divine guidance.

AlBe: can we get to some more recent examples of hidden (or not So hidden) histories, as most of us are familiar with the speculation concerning the hidden history of 'the Wandering who,' and their Kazarian kinship from old days when the dark sorcerers of those lands converted, and then enjoyed, at their whim, the Talamudic techniques of dealing with the peasant goyim (as is parodied and presented 'in your face' in those 'protocols Of the elders of zion').

Fate: well, the wandering sOb's did kill off the shaman Freud - one of their tribe mind you - for daring to write a book (his last book) about the true history of his tribe, starting with the Hyskos invasion (by way of deception) of Egypt, and their false King's effort at changing the religion of the land, and various efforts aimed at stealing their theurgy techniques, like in, the now corrupted, 'Hiram Abif legend'

god And devil

> *recorded history is so problematic, for reasons already discussed by the interlocutors, but for many other reasons also. A lasting legacy of false history however can also include subtle mind control techniques that can, for example, twist the nature of what is human nature*

Sophia: the Hebrew god entity, the Demon Jehovah, that developed in those subverted writings played both roles of what some came to call god And devil.

herb: is not the fable of Abraham, for example, also a corruption of an ancient time when humans (and you k-now who you are) were being assisted by the 'alien man,' who informed the humans (and you k-now who you are) that they did not have to sacrifice their children to the ruling 'god.'

Sophia: indeed my dear herb, as the ancient Baal, worshiped in those days, was a (mountain) god of the Hyskos, who while credited with many wonders, was a 'god' whom demanded much of its subjects.

The priest class most certainly can be labeled as sons of Beliel (those sOb's), and they, as a matter of course, always desired blood sacrifices.

Fate: and these sacrifices continued on in the historical records in the form of wars, for you must see and please understand; wars are not an inherent part of human nature, but nevertheless, needless wars became the favored sacrificial techniques of the sOb's.

herb: the sons of the Serpent became active in long ago times after a great cataclysm, and agreed to make efforts to free the 'Mind' of the newest brand of slaves of the sOb's, and even though stories of their struggles remain, the stories are always turned inside out by the sOb's.

Sophia: indeed my dear herb, and this is also cleverly done in the control manual (the KJV) put out by the sorcerers Dee and Bacon, wherein there is, for example, a general (or major) prohibition against k-nowledge, and restriction on any type of magical practices whatsoever.

herb: and it was under a threat of death that a peasant would dare pay heed to someone who would dare offer a different view on 'things' than offered up - in those times - by the sOb's. Curious, is it not, that all our (false) historical relgous records call for, above all else, obedience, and always warn against the acquisition of k-nowledge.

Fate: and this technique continued on after the original control manuals were developed, wherein the sOb's went and formulated a complete mythos around an idea of a satan; with the help of the magus Milton for example, and thus exert pressure on the peasants to oppose anything that would oppose the established old World order (of those sOb's).

Sophia: and the KJV control manual also prohibited masturbation.
herb: da heck you say?

and Wo/Man's relationship to it

Joe and Giordano wince a bit, especially Joe, and they both ponder a bit as to how this relates to the topic 'on What is'

Giordano: my dear herb... I k-now that it is important to understand the true(r) history of things, but are all these asides really germane to the grand efforts we set to ourselves of trying to understand / contemplate 'on What is'?

herb: my dear Giordano, we can consider some of this un-covering of our hidden history as related to our Pythagorean basics of 'understanding the nature of universe and man's relationship to it.' Sadly, if we only have experienced thoughts relating to a false-history, then we are somewhat limited in moving our thoughts forward... which is to say they (those thoughts - are thoughts alive?) may be mostly deceiving thoughts, having as a basis they do in a false / hidden history.

Joe: so, my dear herb, even though such 'real' thoughts could have someone put in jail (or to the sword) in certain countries in certain times, you do feel it important that we understand hidden history, in order to understand 'man's relation to it'.

Sophia: where the 'it' must be the Universe, and all the people in it, eh, my dear herb?

patton Must die

> *at this sequence of moments, it is not only herb who feels compelled to wander a bit amongst the ruins, some more aware than others of the theurgetically charged statues and pathways, and then (the goddess) Sophia watches as Giordano finds laying about a copy of the ancient freedom enchantment entitled '1984'*

Sophia: the great shaman Orwell famously quipped:

> In a time of deceit, telling the truth is a revolutionary act.
> - George Orwell

and, as AlBe intuited, there are a many many immortal souls whom had a particular incarnation in the / this physical realm seemingly cut short after offending, or for not obeying, the sons of Beliel (those sOb's).

herb: a really obvious example of this type of offending is what happened to the shaman General Patton at the end of the so-called World War II.

Sophia: indeed, and even though it had something to do with a 'lie so big that it can not be mentioned'

which, when considering this group of interlocutors, must be really big

the general was simply not willing to go along with the remainder of the war plans.

AlBe: and while we will not ask you dear (the goddess) Sophia to discuss the 'lie so big that it can not be mentioned,' can you tell some of the other war plans that the shaman Patton may have been opposed to?

Fate: if I may my dear (the goddess) Sophia, I want to mention how the shaman Patton was indeed a colleague of the great shaman Smedley Butler, and thus he was well aware of the depth of the tentacles that the sOb's had into the hegemonic country of his birth during these times.

For now, we should point out that the shaman Patton was not going to sit quietly by while *Operation Paperclip* was put into action, unless, that is, it was made public, as you see, the shaman Patton had great respect for the German soldier. Nor was this shaman willing to allow and/or oversee the wanton slaughter of German civilians that was planned, and eventually carried out by the sOb's and the untold number of Bolshevik servants at their command; and thus, it was determined (and apparently approved) that patton Must die.

satan Came, satan saw, and a civilization falls into ruins

> *as the interlocutors are leaving the ancient ruins, herb feels nudged to recount another atrocity from those times when the sons of Beliel still ran rampant, a story by the shaman J. Speer-Williams - told with his permission - by stand-in interlocutors: the demon sorceress Hillsfury (played by Glia), the shaman Jack (played by thalamus), and herb proudly playing the part of the Colonel - as all of us should accept a k-nowledge of our fate*

Hillsfury:

> Ha-ha-ha-ha-ha
> Ha-ha-ha-ha
> Ha-ha-ha-
> Ha-ha
> Ha!

Jack: Not so long ago, in a far away land there lived a benign ruler of a diverse people, a people who had little history of being a territorially distinct nation. For many of the people in this country, their nation-state allegiances ended at the boundaries of their particular tribe.

The kindly ruler's problems were exacerbated by the sheer size of his country - a whopping 1,760,000 square kilometers - and its disparate and aggressively autonomous tribal groupings were of no help either. How could anyone form these huge cultural differences into one cohesive nation?

In this hot dry land of mostly desert waste was the additional problem of food, which was always at a premium. More important yet was water, which was as valuable as gold.

How could the head of a country, with such problems, become a benign ruler when most of his people had no sense of national unity and no willingness to come together as one people, one nation?

It is said that when the right man meets the right opportunity, miracles will occur. Such a man appeared: his name was Muammar Gaddafi.

Hillsfury: The opportunity was oil - lots of it.

Jack: The country? It was Libya, in northern Africa on the Mediterranean Sea. If anyone knows of a national leader who shared the wealth of his country's resources with the people of his nation more than Colonial Gaddafi did, please tell me who that man is or was.

Hillsfury: For years, the banker owned and controlled press demonized Muammar Gaddafi; but what he accomplished for his people during that time period seems to be that of storybook legend.

Jack: It has been said, that Colonel Gaddafi was never happier than when he was in the desert, amongst his people, and was always traveling with his Bedouin tent. He was truly a man of his people.

And by operating his country outside the tight financial orbit controlled and owned by the pathological and morbid International Banking and Monetary Cartel, Gaddafi made his country and its people the richest in Africa. As a result, however, he made some very powerful enemies - enemies who will not tolerate any nation to stand on its own and be economically independent.

Hillsfury: It was, however, his plan to exchange his nation's oil for gold (or a gold-backed currency), instead of the inflated US reserve currency that probably prompted his brutal killing.

Jack: Such men as Muammar Gaddafi only come our way once in a century, proof of which is how long he stood up to the most potent, anti-life power on earth, the Rothschild-banking mafia – who control the US/NATO nations.

Hillsfury: But in time, the wolves circled and eliminated him, then blackened his name in history.

and this is Jack's humble effort to give a great man a better eulogy

Jack: Gaddafi's government had its own wholly-owned central bank that issued loans to its people free of interest, as riba (usury) was not permitted in Libya at that time. This was, of course, serious sacrilege to the usurers of Earth and it peoples.

Beyond the nomadic Bedouin and Tuareg tribes, most Libyan families owned both a house and a car. Again, Colonel Gaddafi had to be eliminated as the banking cartel considers middle classes anywhere as threats to their power.

Hillsfury: Before the fifty-plus daily US/NATO bombings of Libya, and its defenseless civilian populations began, Muammar Gaddafi's government gave everyone free healthcare and education. Libyans enjoyed a literacy rate of over eighty percent.

Literacy and an education that teaches critical thinking was also a threat to the power structure. Had Colonel Gaddafi followed the American model of governmentally forced stimulus-response education of teaching children what to think instead of how to think, he might have bought himself a few more years of life.

Jack: Before the bombings, largely of American origin, life expectancy in Libya was seventy-five years, the highest in Africa, and about ten percent above the world average.

Preceding the US-led air attacks, there was little to no unemployment in Libya.

Hillsfury: Note: The United States of America has long been under the heavy control of the International Banking/Monetary Cartel; and as a result, today, we have, at least, 92 million unemployed citizens.

Jack: One of the biggest lies ever promoted by the corporate press is 'War is good for the economy.'

War destroys economies due to the extensive debt-based usury lending involved. Strong economies are made by producing in abundance needed and wanted goods and services that are sold at home and around the world.

Production, not destruction, builds a healthy economy.

Under Gaddafi, Libya gained the highest gross domestic product in Africa, with less than five percent of its population classified as poor.

Libya, a hot, dry, dusty country, long suffered with a lack of clean water. Under the leadership of Muammar Gaddafi, however, the world's largest infrastructure project was completed. It was a man-made, underground river, undoubtedly made from what is now known as Primary Water. The project provided clean drinking water to seventy percent of the Libyan people. It held the potential of turning vast wastelands into farmlands - that is, until US/NATO bombs destroyed its pumping stations, thus destroying the entire project.

Hillsfury: The power structure had to destroy Gaddafi's Man-made River Project, before it told the world that never again would man have to be without clean water to drink or to farm his deserts. They would not allow Gaddafi to spoil the coming water-wars that were on their drawing boards.

Jack: In 1991, at the gala opening of the Man-made River Project, the good Colonel Gaddafi spoke to the invited dignitaries and assembled crowd:

Colonel: "After this achievement, American threats against Libya will double. The United States will make excuses [but] the real reason is to stop this achievement, to keep the people of Libya oppressed."

The Colonel's words were prophetic

Hillsfury: After American super bombers laid waste to Libya's water and electricity, the bombers destroyed the country's main food supply.

Jack: For many decades, beginning in the summer, thousands of camels made a three-month journey from the grass-lands of Libya to metropolitan markets.

Amid the horrifying wailing and whimpering from thousands of camels and herdsmen being mindlessly and thoroughly slaughtered, the US/NATO bombs fell until animal and human flesh were ready to rot in the hot Libyan deserts.

For months and years the American mainstream media had demonized Muammar Gaddafi, often saying the goal was to bring democracy to Libya by removing their dictator.

So how did the US government justify the barbarity of butchering thousands of camels, while people were starving all over the rest of Africa?

Perhaps you heard the absurd answer from your own television set:

Hillsfury: The camels were carrying weaponry to support those loyal to Colonel Muammar Gaddafi.

Jack: Sadly many Americans were happy with that answer never realizing that certainly, during the dark of nights, Libyan troops could have quickly transported their weapons all over their country in a matter of days, not months. Death and destruction of all life forms are the hallmarks of the military forces controlled by the foreign Banking and Monetary Cartel.

Hillsfury: The awful destruction of the once free and independent Libya should serve us as a moral lesson, representative of our atrocities in Iraq, Afghanistan, Syria, Lebanon, and Palestine, where millions of people have been killed, injured, or made homeless in a radioactive hotbed of chaotic waste, with no known way of cleaning it up.

For no military tactical or strategic reasons, depleting uranium (DU) has been employed in all of our bombings in the Middle East.

But still, the Western press calls our insane international crimes against humanity humanitarian interventions.

When in fact, the horrific US/NATO bombing of Libya was a banker's looting operation for oil, water, gold, and to bring a brave and recalcitrant country into the tightly controlled orbit of usury finance and economic slavery.

Jack: I now ask my brothers and sisters in America and Europe to join with me in a prayer that the US/NATO axis of evil destruction desist from any more such humanitarian interventions.

a Sekret history of amerika

while the interlocutors whisper among themselves about how history does show itself sometimes, Giordano asks to hear about another hero that simply had to die

Giordano: and this was not an uncommon occurrence in historical times, in that many a person who was a hero at one moment, became persona non gratis a moment or two after.

herb: do you have someone in mind my dear Giordano?

Giordano: indeed my dear herb, as I am wondering what more there is to tell about those explorers Lewis and Clark, where only one of them was allowed to live.

thalamus: if I may dear herb, as I have pondered deeply about these times in the hegemonic country of your birth. The explorations of Lewis and Clark, you see, proceeded the deliberate attempted genocide of all Native American cultures through the machinations of the sons of Beliel (those sOb's).

Giordano: but why?

Fate: the sOb's did not want, and could not allow (in their Archon infested minds), for an accurate historical account of these lands - i.e., the hegemonic country of herbs birth - to be made public k-nowledge. Because, as you see, the explorers encountered all kinds of evidence as it concerned ancient civilizations (and worse of all, some concerned the sons of the Serpent).

Glia: is this related to that story of the large treasure and cave system found in the Grand Canyon that was hushed up by the Smithsonian Institute?

Fate: that and many many other things my dear Glia. And while many things found by the explorers could be destroyed or buried, the sOb's of course worried tremendously about word getting out of an ancient historical past when interactions between the newcomers and the native tribes would become more frequent.

So, as the sOb's did in the South America conquest, genocide and the burning of evidence was the preferred choice for what remained. Along with these tried and true methods of the sOb's, a new experiment in mind control was also started: herein the sOb's would effectively steal all the children of the red man, and then provide a different type of 'education' for them than would have been provided in their normal historical upbringings (which of course involved good historical stories of historical times). Other astrocities at the whim of the sOb's would also be conducted, because as one should always remember, those sOb's are always looking for ways to perform blood sacrifices for their ever present demiurge.

herb: and thus one of Lewis and Clark must die?

Fate: yes, one had to die.

war Is a racket

> *the interlocutors are exhausted, for while it is important to understand the true way of ancient proceedings, wiping away ones tears, and hearing those ancient cries, well ...*

herb: before we retire for the evening, let us gather by the fire, wipe our tears, and together agree to harbor no fear.

Giordano: now, my dear herb, we have discussed much about hidden history, all as part of our continuing efforts at understanding the Pythagorean basics required in order to better approach the topic 'on What is'. However, does the heroes journey[98] always require that the hero gets killed off by the sons of Beliel (those sOb's)? is there no other option but death for the hero?

herb: there is the case of the shaman General Smedley Butler, and while it may be the case that he had to get permission from the sOb's to speak plainly, as did, you may recall, that tool of the sOb's, General President Eisenhower, when he gave his famous warning about the 'military industrial complex.'

Giordano: and yes, was it not the shaman General Smedley Butler who wrote the book 'War is a Racket'?

herb: yes my dear Giordano, and in that famous book, which became rather hard to find, the shaman General Smedley Butler wrote how he became rather ashamed of his doings when he realized he was only a tool / strong arm for his corporate masters (those sOb's). When ransacking and terrorizing the peoples of Central and South America as part of his job in the military of the hegemonic country of his birth, he finally saw his actions for what they were.

However, what is more important, is what transpired before he wrote that particular freedom enchantment.

Sophia: indeed, my dear herb, because as it is, the shaman General Smedley Butler is maybe one of the few who was able to say 'no' to the sons of Beliel (those same sOb's) and live to talk about it.

herb: indeed, and further, he gave testimony in front of Congress about how he was approached by an influential group of banksters and other associated conspirators to lead a fascist coup to take over the government of Amerika.

Fate: but as Fate would have it my dear herb, the records of his important congressional testimony, which has copies traditionally spread about to many libraries around the country, soon began to disappear from the archives of said libraries, with hardly any to be found now anywhere.

---

[98] explained well by the shaman Joseph Campbell and employed in almost every script of those famous Holly-wood movies of old.

not So hidden

> *it has been a full cycle of the moon since the interlocutors existed the enchanted forest via the ancient river Eridanus in rafts that herb and AlBe had placed there almost 4 moons prior. The interlocutors have agreed to meet again at one of the many week long music shows that occasionally pop up here or there; and while Giordano and the goddesses Sophia and Fate are yet to arrive, the interlocutors are together around a large fire where others are also ready to lend their attention*

thalamus: my dear herb, many of the magicians musicians that have gathered are familiar with your perchant for hidden history, and some are asking for some discussions as it concerns wars, banking, and spies.

herb: and why my dear thalamus, are the truths about wars, banking and spies all cubbied away in hidden history?

Glia: as you k-now well my dear herb, there are some aspects of hidden history that are not So hidden. To wit, what can you tell us about the history of the Russian lands when they were taken over (by way of deception) by the sons of Beliel (those sOb's)?

herb: many thanks my dear Glia, as you are totally correct when you mention that this particular endeavor by the sOb's was not So hidden, for it is well k-nown that the so-called revolution was financed from London and Amerika. Butt to set the stage properly, we all here about the fire should take a deep 3 breaths and listen closely to these not So hidden historical words attributed to the great shaman Alexandr Solzhenitsyn:

> You must understand, the leading Bolsheviks who took over Russia were not Russians. They hated Russians. They hated Christians. Driven by ethnic hatred they tortured and slaughtered millions of Russians without a

shred of human remorse. It cannot be overstated. Bolshevism committed the greatest human slaughter of all time. The fact that most of the world is ignorant and uncaring about this enormous crime is proof that the global media is in the hands of the perpetrators.
- Alexandr Solzhenitsyn

judea Declares war

many about the fire k-now the depths (or at least the shallow waters) of depravity of the Bolsheviks, but the memory hole has swallowed another related part of the story, which is shared by (the goddess) Fate, who recently snuggled up to the camp fire

Fate: and as Fate would have it, Hitler k-new well about the Bolshevik slaughter, and many of the sons of Beliel (those sOb's) who scattered with the fall of the debauchery that was the Wiemar Republic, fled to the Bolshevik stronghold. Here, they would be welcomed by the demon Stalin, a Georgian, not a Russian, but from that tribe of Khazars who joined the Wandering who way back in order to taste of the Talamudic poison that was most recently distilled in Babylon.

Sophia: greeting my dear interlocutors, and others who would, by their own volition, take on the most Dangerous of all undertakings. Are we here seeking wisdom and freedom (if we are lucky to be so divinely inspired) through discussions and fire tending?

My dear herb, has it been mentioned yet how it was that 'judea Declares war' on Germany was proclaimed from the Bolshevik stronghold on that 24th of March in the so-designated year 1933, a full four (4) days before the reactionary German embargo against said group?

herb: as to your query, no we have not discussed those particular facts, which are not So hidden one should presume, because they were in all the papers of the so mentioned day (butt maybe now expunged from records? if the sOb's had their way).

Fate: and there are many other things about those times, some more hidden than others, as it concerns financing, repatriation, stolen lands, and of course the 'lie so big that it cannot be mentioned;' oh my, it is all so sad. For example, the Stern letter of 1941, that outlined the obvious plan to cooperate with the Germans to establish a homeland; in effect stealing the land, and all the mineral rights, of what was Palestine

Giordano: the Balfour declaration all over again, the sOb's never give up

herb: and now we need step back 20 years before judea Declares war (officially), before the treacheries of the so-called World War I, and speak about the treachery that was committed by the congress of the hegemonic country of my birth on a Christmas Eve, in the absence of a quorum

not So federal reserve

> *a few of the interlocutors are queried by others around the glorious fire pit, that, by the way, is speaking volumes also. (the goddess) Sophia, and (the goddess) Fate, recently arrived, also draw the typical attention given by those who can bear witness to the divine aura, and even by those who only have a hint of its presence*

Giordano: my dear herb, and newly inspired interlocutors, if Fate allows, I would like to introduce the deception that occurred on that fateful Christmas eve of the so-designated year 1913.

Butt before recounting those illegal treacherous deeds, we must go back and bear witness to the truth behind the sinking of the Titanic, for you see, not everyone in those days was ready to accept a private central bank run by the sons of Beliel (those sOb's).

Glia: my dear Giordano, are you referring to that story where many of the industrialists who opposed the creation of a (not So) federal reserve were lured to a meeting on the famous Titanic, only to be abandoned by the organizer, and then drowned at sea?

Giordano: yes my dear Glia, and those who did not perish were - don't you know - duly warned by those sOb's

> *thalamus is waiting to speak before he is to depart and setup a decoy fire pit to help the interlocutors escape the imminent intervention of agent provocateurs*

thalamus: and this same type of deadly threat is seen over and over in those times, earlier times, and later times. Does anyone recall almost a century after these facts we are now discussing, when the sOb's arranged the slaughter of the children of the Norwegian politicians who were set to take a stand against the sOb's?

Giordano: yes, indeed, in much the same fashion, by way of deception or outright aggression, those sOb's are unrelenting.

herb: can we share with these newbie interlocutors, who may so brave, to take on the most Dangerous of all undertakings, here, nOw, the remainder of the story of the treachery of that Christmas eve, that would place the world in virtual bondage for the next century?

Fate: if I may dear Giordano, as Fate would have it, on that fateful Christmas eve, without even enough members in the chamber to take an official vote, the Federal Reserve Act was 'passed' in congress. When everyone returned, not one was successful at stopping the insane legislation that gave a private group of banksters the usurious rights to taxes (also newly instituted were those, and also done illegally, of course) taken from the peasants when those sOb's printed money to 'lend' to amerika.

young interlocker: and not a one among the elected officials reproached this legislation upon their return from break?

agent provocateur: they probably were worried that they and their families would be next to die if they did.

herb: indeed, which is the response expected by the sOb's

when they run such a long-planned coup. Butt, the players involved then, any involved now, and any who would allow destiny and necessity to set their involvement in the future, need to understand that the sOb's are not masters of every immortal soul; and the sorcery of those sOb's, while deliberate and strong, does indeed, the divine willing, have bounds.

war Funding

> *thalamus, Giordano, and Glia have been enlightened by (the goddess) Sophia and (the goddess) Fate that they should follow herb in a shamanic dance to the fire pit that AlBe and Joe have prepared near the tree line. (the goddess) Fate k-nows that agent provocateurs are red-eye to start trouble (and not the game)... (the goddess) Sophia plays anyway*

Sophia: and with the (not So) federal reserve now in charge of money printing, and the collection of demoniacal inspired usury, the stage was set to begin the slaughter (and blood sacrifices) that was a World War

Fate: while wars have never been about what they seem to be, the creation of a new private bank, in the hegemonic country of herb's birth, gave the sons of Beliel (those sOb's) the signal to start, those long planned blood sacrifices and rituals that would be called in historical records both world wars I and II, and usher in the times that the dark Magus John Dee said was needed, and also the time, if you recall, for the sOb's plan which called for silicon crystal mastery

Sophia: and without bankster support, there could / would be no more wars, and no ritual slaghter, for that was not originally part of human (and you k-now who you are) ways, until the rabid Dogs had full sway

Fate: it is always by way of deception, as some of you, here, nOw, about this glorious fire pit, k-now

shadows suggest the interlocutors are now warming about a fire pit near a tree line, but they are slowly making their way through a new forest growth, about an older city that was abandoned, so that no roads would lead there anymore

thalamus: while it was nice to mingle with potential interlocutors and even a would be agent provocateur, I suppose it is time to crack on, so lets see some of this city, that was abandoned long ago
Joe: right.

rabid Dogs have full sway

the interlocutors meet up again about 1/2 of a moon away from when they left (the goddess) Sophia and (the goddess) Fate at a fire pit. They all catch up with each other, while herb and AlBe are fishing for trout, near an olde railroad trestle, that brought back many olde memories no doubt

Joe: when we were talking about hidden (or not So hidden) histories, I thought we should talk again about those groups who purposely, and openly, keep a hidden history amongst themselves.
Of course I speak now of the many many sekret societies, seemingly rampant in recent historical times.
herb: an while being a member of a sekret society should not be considered sorcery in and of itself, it may indeed keep us humans (and you k-now who you are) from k-nowing more of our true history.
AlBe: why is there a need for secret societies; I mean, do they have something to hide?
Sophia: traditionally, if one had an opinion or idea or belief that was outside of the accepted norms of their living situation, one might want to be very careful about whom one shared that opinion with.

AlBe: so then, we cannot or should not say right away if a secret society is good or bad, but maybe assume it started as a way to protect one from the dominant cultural norms?

herb: that is a perfectly good rationale for members to keep to themselves, but we should also suspect that those whom drive, and possibly create cultural norms, they also have a need for sekret societies.

Sophia: in order to keep the true intentions of the sons of Beliel (those sOb's) a sekret from the peasant goyim, or the Roman plebeian, no doubt.

herb: yes my dear (the goddess) Sophia, or to keep things from the sOb's - if that is possible. In the final analysis however, we must / should agree that some information is being kept from us peasants, and the shaman dr. Steven Greer had this to say in his opus 'Hidden - Truth - Forbidden Knowledge':

> We live, unfortunately, in a time where spirituality has been equated with passivity. This is a very dangerous thing, because it's a propaganda or indoctrination that's designed to create people who are spiritually oriented and very nice, but very passive and ineffectual. And in that passivity, the rapid dogs can have full sway.
> - Dr. Steven Greer

which very well surmises some of the problems we face when discussing secret societies, the sOb's, and the grand idea of 'on What is.'

Giordano: and on the topic of sekret societies, lets not forget the advice from the shaman Groucho Marx, who quipped: 'I do not know if I want to join a group that would have me as a member.'

herb: the shaman dr. Steven Greer, when talking this way, was above all else concerned with the phenomena of UFOs that were so prevalent in those times.

Fate: not realizing, as would become clear in time, that the UFO was not of extraterrestrial origin, butt rather right out of Atlantis

an Atlantis of plato

> *the interlocutors laugh and laugh (and laugh) in response to (the goddess) Fate's reference to Atlantis, and before the great topic was continued, a meal that included chicken-fried stream trout was served up*

herb: the Atlantian mythos, in either the shaman Manly P. Hall's '*On the Secret Teaching of All Ages*,' or by way of Plato and other ancient accounts (including his Egyptian sources) of an ancient race / ancient time on the planet is well k-nown.

Sophia: thoth Hermes Trismegistus was a theurgist of Atlantis, and in those times, it probably meant him a rebel and outcast - as he eschewed sorcery (which was the dominant magikal style in the later days of Atlantis).

herb: thoth had also written over 23,000 books over the ages, because as you see, he had mastered the navigation of the immortal soul between the realms, and could retain memories between each incarnation in the / this physical realm of the demiurge.

Sophia: and thus the great shaman thoth was on more than equal footing with my sun, and in general, could not be troubled by the Archons.

Joe: was Atlantis always a sorcerers paradise, or did it just end up that way?

herb: before the first revolt that saw the sons of the Serpent depart Atlantis, the occupants and their manufactured prodigy got on well together. Millennium, combined with toned k-nowledge of the invisible workings of the Universe (including mastery of that mysterious fluid-type magical substance called electricity), gave the opportunity to some of the priests to choose sorcery over theurgy.

Fate: many say that the change was due to one of the first invasions, by way of deception, wherein a principle target of the invasion were the mysteries understood there, and the celebrated, and now maybe lost, 'Book of Thoth[99].'

---

[99] from 'The Secret Teaching of All Ages,' "this work contained the secret processes by which the regeneration of humanity was to be accomplished..." and "its pages were covered with strange hieroglyphic figures and symbols, which gave those acquainted with their use unlimited power over the spirits of the air and the subterranean divinities."

thoth And poimandres

> *herb and Giordano agree to perform 'The Divine Pymander,' with thoth / Hermes (played by Giordano), and the Great Dragon / poimandres (played by herb). Herethen, a 'Vision of Hermes,' told throughout the ages, translated from Arabic And Greek in the so-designated year 1650, and also told recently by Manly P. Hall ...*
>
> *thoth / Hermes, while wandering in a rocky and desolate place, gave himself over to meditation and prayer. Following the secret instructions of the temple, he gradually freed his higher consciousness from the bondage of his bodily senses; and thus released, his divine nature revealed to him the mysteries of the transcendental spheres. He beheld a figure, terrible and awe-inspiring. It was the Great Dragon*

Great Dragon: Hermes, why have you meditated upon the World Mystery?

Hermes: will you tell me whom you are?

Great Dragon: poimandres, the Mind of the Universe, the Creative Intelligence, and the Absolute Emperor of all.

Hermes: can you tell me the nature of the Universe, and the constitution of the gods?

> *the Great Dragon acquiesced, bidding thoth hold its image in his mind. Immediately the form of poimandres changed. Where it had stood there was a glorious and pulsating Radiance. This Light was the spiritual nature of the Great Dragon itself. Hermes was raised into the midst of the Divine Effulgence and the Universe of material things faded from his consciousness.*

> *Presently a great darkness descended and, expanding, swallowed up the Light. About Hermes swirled a mysterious watery substance which gave forth a smokelike vapor. The air was filled with inarticulate moanings and sighings which seem to come from the Light swallowed up in the darkness. His mind told Hermes that the Light was the form of the spiritual Universe and that the swirling darkness which had engulfed it represented material substance.*
>
> *then out of the imprisoned Light a mysterious and Holy Word came forth and took its stand upon the smoking waters. This Word - the Voice of the Light - rose out of the darkness as a great pillar, and the fire and the air followed after it, but the earth and the water remained unmoved below. From the waters of Light were formed the worlds above and from the waters of darkness were formed the worlds below*

poimandres: i thy God am the Light and the Mind which was before substance was divided from spirit and darkness from Light. And the Word which appeard as a pillar of flame out of the darkness is the Son of God, born of the mystery of the Mind. The name of that Word is 'Reason.' Reason is the offspring of Thought and Reason shall divide the Light from the darkness and establish Truth in the midst of the waters. Understand O Hermes, and meditate deeply upon the mystery. That which in you sees and hears is not of the earth, but is the Word of God incarnate.

> *So it is said that Divine Light dwells in the midst of mortal darkness, and ignorance cannot divide them. The union of the Word and the Mind produces that mystery which is called 'Life'*

poimandres: As the darkness without you is divided against itself, so the darkness within you is likewise divided. The Light and fire which rise are the divine man, ascending the path of the Word, and that which fails to ascend is the mortal man, which may not partake of immortality. Learn deeply of the Mind and its mystery, for therein lies the secret of immortality.

> *the Dragon again revealed its form to Hermes, and for a long time the two looked steadfastly one upon the other, eye to eye, so that Hermes trembled before the gaze of poimandres. At the Word of the Dragon the heavens opened and the innumerable Light Powers were revealed, soaring through Cosmos on pinions of streaming fire. Hermes beheld the spirits of the stars, the celestials controlling the Universe, and all those Powers which shine with the radiance of the One Fire - the glory of the Sovereign Mind. Hermes realized that the sight which he beheld was revealed to him only because poimandres had spoken a Word. The Word was Reason, and by the Reason of the Word invisible things were made manifest. Divine Mind - the Dragon - continued its discourse*

poimandres: Before the visible Universe was formed its mold was cast. This mold was called the 'Archetype,' and this Archetype was in the Supreme Mind long before the process of creation began. Beholding the Archetypes, the Supreme Mind became enamored with Its own thought; so, taking the Word as a mighty hammer, It gouged out caverns in primordial space and cast the form of the spheres in the Archetypal mold, at the same time sowing in the newly fashioned bodies the seeds of living things. The darkness below, receiving the hammer of the Word, was fashioned into an orderly Universe. The elements separated into strata and each brought forth living creatures. The Supreme Being - the Mind - male and female, brought forth the Word; and the Word, suspended between Light and darkness, was delivered of another Mind called the 'Workman,' the 'Master-Builder,' or the 'Maker of Things.'

In this manner it was accomplished, O Hermes: The Word moving like a breath through space called forth the 'Fire' by the friction of its motion. Therefore, the Fire is called the 'Son of Striving.' The Workman passed as a whirlwind through the Universe, causing substances to vibrate and glow with its friction. The Son of Striving thus formed 'Seven Governors,' the Spirits of the Planets, whose orbits bounded the world; and the Seven Governors controlled the world by the mysterious power called 'Destiny' given them by the Fiery Workman. When the 'Second Mind' (The Workman) had organized Chaos, the Word of God rose straightaway out of its prison of substance, leaving the elements without Reason, and joined Itself to the nature of the Fiery Workman. Then the Second Mind, together with the risen Word, established Itself in the midst of the Universe and whirled the wheels of the Celestial Powers. This shall continue from an infinite beginning to an infinite end, for the beginning and the ending are in the same place and state.

Then the downward-turned and unreasoning elements brought forth creatures without Reason. Substance could not bestow Reason, for Reason had ascended out of it. The air produced flying things and the waters such as swim. The earth conceived strange four-footed and creeping beasts, dragons, composite demons, and grotesque monsters. Then the Father - the Supreme Mind - being Light and Life, fashioned a glorious Universal Man in Its own image, not an earthly man but a heavenly Man dwelling in the Light of God. The 'Supreme Mind' loved the Man It had fashioned and delivered to Him the control of the creatures and workmanships.

The Man, desiring to labor, took up His abode in the sphere of generation and observed the works of His brother - the Second Mind - which sat upon the Ring of Fire. And having beheld the achievements of the Fiery Workman, He willed also to make things, and His Father gave permission. The Seven Governors, of whose power He partook, rejoiced and each gave the Man a share of Its own nature.

The Man longed to pierce the circumference of the circles and understand the mystery of Him who sat upon the Eternal Fire. Having already all power, He stooped down and peeped through the seven Harmonies and, breaking through the strength of the circles, made Himself manifest to Nature stretched out below. The Man, looking into the depths, smiled, for He beheld a shadow upon the earth and a likeness mirrored in the waters, which shadow and likeness were a reflection of Himself. The Man fell in love with His own shadow and desired to descend into it. Coincident with the desire, the Intelligent Thing united Itself with the unreasoning image and shape.

Nature, beholding the descent, wrapped herself around the Man whom she loved, and the two were mingled. For this reason, earthly man is composite. Within him is the Sky Man, immortal and beautiful; without is Nature, mortal and destructible. Thus, suffering is the result of the Immortal Man's falling in love with His shadow and giving up Reality to dwell in the darkness of illusion; for being immortal, man has the power of the Seven Governors - also the Life, the Light, and the Word - but being mortal, he is controlled by the Rings of the Governors - Fate or Destiny.

Of the Immortal Man it should be said that He is hermaphrodite, or male and female, and eternally watchful. He neither slumbers nor sleeps, and is governed by a Father also both male and female, and ever watchful. Such is the mystery kept hidden to this day, for nature, being mingled in marriage with the Sky Man, brought forth a wonder most wonderful - seven men, all hermaphrodite, both male and female, and upright in stature, each one exemplifying the natures of the Seven Governors. These, O Hermes, are the seven races, species, and wheels.

After this manner were the seven men generated. Earth was the female element and water the male element, and from the fire and the aether they received their spirits, and Nature produced bodies after the specied and shapes of men. And man received the Life and Light of the Great Dragon, and of the Life was made his Soul and of the Light his Mind. And so, all these composite creatures containing immortality, but partaking of mortality, continued in this state for the duration of a period. They reproduced themselves out of themselves, for each was male and female. But at the end of the period the knot of Destiny was untied by the will of God and the bond of all things loosened.

Then all living creatures, including man, which had been hermaphroditical, were separated, the males being set apart by themselves and the females likewise, according to the dictates of Reason.

Then God spoke to the Holy Word within the soul of all things, saying: "Increase in increasing and multiply in multitudes, all you, my creatures and workmanships. Let him that is endued with Mind know himself to be immortal and that the cause of death is the love of the body; and let him learn all things that are, for he who has recognized himself enters into the state of Good."

And when God had said this, Providence, with the aid of the Seven Governors and Harmony, brought the sexes together, making the mixtures and establishing the generations, and all things were multiplied according to their kind. He who through error of attachment loves his body, abides wandering in darkness, sensible and suffering the things of death, but he who realizes that the body is but a tomb of his soul, rises to immortality.

thoth: but why should men be deprived of immortality for the sin of ignorance alone?

poimandres: to the ignorant the body is supreme and they are incapable of realizing the immortality that is within them. Knowing only the body which is subject to death, they believe in death because they worship that substance which is the cause and reality of death.

thoth: how then can the righteous and wise pass these 'tests' and have communion with God?

poimandres: That which the Word of God said, say I: "Because the Father of all things consists of Life and Light, whereof man is made. If, therefore, a man shall learn and understand the nature of Life and Light, then he shall pass into the eternity of Life and Light."

thoth: what then is the road to this knowledge of Life and Light, and thus Life eternal?

poimandres: Let the man endued with a Mind mark, consider, and learn of himself, and with the power of his Mind divide himself from his not-self and become a servant of Reality.

thoth: do all men have Minds, and are all capable of understanding what you say here to me?

poimandres: Take heed what you say, for I am the Mind - the Eternal Teacher. I am the Father of the 'Word' - the Redeemer of all men - and in the nature of the wise the Word takes flesh. By means of the Word, the world is saved. I 'Thought' (Thoth) - the Father of the Word, the Mind - come only unto men that are holy and good, pure and merciful, and that live piously and religiously, and my presence is an inspiration and a help to them, for when O come they immediately know all things and adore the Universal Father. Before such wise and philosophical ones die, they learn to renounce their senses, knowing that these are the enemies of their immortal souls.

I will not permit the evil senses to control the bodies of those who love me, nor will I allow evil emotions and evil thoughts to enter them. I become as a porter or doorkeeper, and shut out evil, protecting the wise from their own lower nature. But to the wicked, the envious and covetous, I come not, for such cannot understand the mysteries of the 'Mind;' therefore, I am unwelcome. I leave them to the avenging demon that they are making in their own souls, for each evil deed adds to the evil deeds that are gone before until finally evil destroys itself. The punishment of desire is the agony of unfulfillment.

> *thoth bowed his head in thankfulness to the Great Dragon who had taught him so much, and begged to hear more concerning the ultimate nature of the human soul*

thoth: what then becomes of our human souls?

poimandres: at death the material body of man is returned to the elements from which it came, and the invisible divine man ascends to the source from whence he came, namely the 'Eighth Sphere.' The evil passes to the dwelling place of the demon, and the senses, feelings, desires, and body passions return to their source, namely the Seven Governors, whose natures in the lower realm destroy but in the invisible spiritual man give life.

After the lower nature has returned to the brutishness, the higher struggles again to regain it spiritual estate. It ascends the seven Rings upon which sit the Seven Governors and returns to each their lower powers in this manner: Upon the first ring sits the Moon, and to it is returned the ability to increase and diminish. Upon the second ring sits Mercury, and to it are returned machinations, deceit, and craftiness. Upon the third ring sits Tiamat, and to it are returned lusts and passions. Upon the fourth ring sits the Sun, and to this Lord are returned ambitions. Upon the fifth ring sits Mars, and to it are returned rashness and profane boldness. Upon the sixth ring sits Jupiter, and to it are returned the sense of accumulation and riches. And upon the seventh ring sits Saturn, at the Gate of Chaos, and to it are returned falsehood and evil plotting.

Then, being naked of all the accumulations of the seven Rings, the soul comes to the Eighth Sphere, namely the ring of the fixed stars. Here, freed of all illusion, it dwells in the Light and sings praises to the Father in a voice which only the pure of spirit may understand. Behold, O Hermes, there is a great mystery in the Eighth Sphere, for the Milky Way is the ground of souls, and from it they drop into the Rings, and to the Milky Way they return again from the wheels of Saturn. But some cannot climb the seven-runged ladder of the Rings. So they wander in darkness below and are swept into eternity with the illusion of sense and earthiness.

The path to immortality is hard, and only a few find it. The rest await the Great Day when the wheels of the Universe shall be stopped and the immortal sparks shall escape from the sheaths of substance. Woe unto those who wait, for they must return again, unconscious and unknowing, to the seed-ground of stars, and await a new beginning. Those who are saved by the light of the mystery which I have revealed unto you, O Hermes, and which I now bid you to establish among men, shall return again to the Father who dwelleth in the White Light, and shall deliver themselves up to the Light and shall be absorbed into the Light, and in the Light they shall become Powers in God. This is the way of the 'Good' and is revealed only to them that have wisdom.

Blessed art thou, O Son of Light, to whom of all men, I, poimandres, the Light of the World, have revealed myself. I order you to go forth, to become a guide to those who wander in darkness, that all men within whom dwells the spirit of 'My Mind' (The Universal Mind) may be saved by My Mind in you, which shall call forth My Mind in them. Establish My Mysteries and they shall not fail from the earth, for I am the Mind of the Mysteries and until Mind fails (which is never) my Mysteries cannot fail.

> *with these parting words, poimandres, radiant with celestial light, vanished, mingling with the powers of the heavens. Raising his eyes unto the heavens, Hermes / thoth blessed the Father of All Things and consecrated his life to the service of the Great Light*

a Great flood

> *the interlocutors enjoy a 'rock and roll' show while they eat, drink and are merry; as the divine shown that day, and surely the next day would bring what it may. The conversation turns back to Atlantis...*

Giordano: and that time in Atlantis that you mention my dear herb, it was before the great battle involving the water goddess Tiamat was it not?

herb: indeed my dear Giordano, as in those times, our planet did not have the great oceans of today, as witnessed by the ancient submerged civilization records, and the extensive labyrinth of tunnels and underground cities.

Sophia: when Tiamat was smashed, some of the remnants became the asteroid belt, but much of the oceans (and octopi and river rocks) that blessed the water goddess Tiamat, eventually fell to this planet, over time, and this is the recorded deluge that is found in all (altered or unaltered) historical memories of all cultures.

Fate: and remnants of the cabal of maniac magicians, who would tutor the sons of Beliel (those sOb's), that were still in Atlantis scattered to various parts of the planet, and to their underground bases also. While the great shaman thoth (Hermes) set about educating the sons of the Serpent where they still survived, you see, it was the great shaman who educated them in the ways of language, writing, metallurgy, civilization, and theurgy; as has always been ack-nowledged in one way or another in all (altered or unaltered) historical memories of all cultures.

Giordano: so what does all this Atlantis stuff have to do with the UFOs of our recent historical times?

herb: you see my dear Giordano, the so-called Atlantians remaining after the flood (sans the sons of the Serpent), already had subterranean accommodations, and many types of spacecraft; as you may or may not k-now, they first came to this planet as outlaws on the run from another star system.

AlBe: so they, the sOb's, or their masters, control the UFOs?

herb: yes, for the most part, this cabal of maniac magicians remain the leading lights (hehe, 'dark lights') for the sOb's.

thalamus: and that hole in the ground found by the shaman Admiral Byrd, that they made him stop talking about, was that an access point to an inner sanctum in the earth?

herb: quite possibly my dear thalamus.

thalamus: and the great pyramids in Antarctica should be included no doubt

AlBe: well then, this makes a lot of sense. So that all the ancient alien talk / propaganda was always meant to throw us humans (and you k-now who you are) off the scent (so to say), so that when the sOb's brought the big holographic show in the sky to the world, it was not a benevolent race of aliens here to help us humans (and you k-now who you are), but it was really just a case of 'meet the new boss, same as the old boss.'

the Invader venus

while the interlocutors seem intent on seeing how far the Atlantaen influences permeates the present day(s), herb will talk briefly about the invader Venus, source of another great cataclysm, before leaving hidden history behind, and moving on to discuss the conspiracy of the 'crisis Of now'

herb: there was another great cataclysm on the planet after the Great flood, one where it is told that the earth stood still for days while a great intruder rambled by on its way to establish a position in the inner sanctum of our solar system, for you see, before this time, earth was the second stone from the son.

Sophia: Jove sent Venus to put pause to the resurgence of the sons of Beliel (those sOb's) on the planet, and at the same time change to calendar from 360 days to 365 days, as the earth had different orbital and rotational characteristics before these events.

herb: and it is strange how ancient astronomer cults all of a sudden became obsessed with (the invader) Venus, the morning star, because as it was, older records show no fascination, or recognition, of this planet.

Fate: the goddess Venus also, during its passings, laid down layers of hydrocarbons onto the earth. These layers of hydrocarbons, stripped right from the goddess, would become the oil fields on this planet, now the third stone from the sun.

Glia: and of course, the use of that gift from the goddess Venus was hijacked by the sOb's, and used as part of their control system in our most recent historical times.

Sophia: indeed, but is not 'oil' also part of the 'hint' parade; in that many were so willing to accept the explanation that 'oil' is a result of the decomposition of old reptiles from the past? How silly was that, because if one is willing to believe a scientist when they say that, then one is willing to believe almost anything.

herb: and when one considers other aspects associated the goddess Venus, specifically its orbital and rotational aspects, we find more songs for the 'hint' parade.

AlBe: what are the orbital characteristics of the morning star?

herb: it takes Venus about 225 earth days to orbit the son, and it takes Venus 117 earth days to rotate once completely about its axis[100]. Thus, a Venus year contains only 1.923, or about 2 Venus 'day's every Venus 'year' (compared to about 365 earth days each earth year, and 670 Mars days each Mars year[101]).

AlBe: and why, my dear herb, is not some of these ancient aspects of our solar system history not taught to all young people?

Fate: consider the fate of the shaman Velikousky if you might my dear AlBe, and you will understand how anyone with ideas / information different than the sOb's propaganda is treated.

Sophia: and why would the sOb's want to teach the peasant goyim anything anyway, for the time had passed when the peasants were actually needed to help master the crystal silicon, and also create for the sOb's the robotic slave machines which could replace, what they typically, and not so affectionately, call the 'useless eaters' - meaning, of course, you peasants.

crisis Of now

> *as the firepit glow grows lower (owowow), herb is intent on bringing up a new topic before the interlocutors retire for the night to work on their individual dreamworld navigation techniques*

---

[100] this means one Venus 'day' is equivalent to 117 earth 'day's.
[101] where a Mars day is about 24.666 earth hours

herb: the conspiracy of the crisis Of now is one of the many tools used by the sons of Beliel (those sOb's) to hijack and subvert the mental processes of us humans (and you k-now who you are).

Giordano: butt my dear herb, is not attention to the 'now' also one of the key techniques used by shamans, magicians and theurgists?

herb: indeed my dear Giordano, butt that 'nOw' k-nown to the theurgists is the 'nOw' that connects the ever present with the ever past and the ever future, a type of non-temporal awareness (or experiencing) technique.

This non-temporal 'nOw' awareness employed by shamans allows for, among other things, divination of future happenings, and connections with other life operating at different temporal scales.

thalamus: eye have heard said that plants and trees do indeed have mental / mind-type activities, butt that they do not 'see' things at the same 10 Hertz rate[102] as us humans (and you k-now who you are).

Joe: right.

Giordano: so what then do you refer to my dear herb when you speak of the conspiracy of the crisis Of now?

herb: it is natural for humans (and you k-now who you are) to contemplate things, and contemplation requires reflection and projection. Meaning, a human (and you k-now who you are) is naturally inclined to contemplate the transcendental and the divine, and search for their true individual history, their individual future, and the current well-being of their immediate family and tribe.

It is during this type of typical contemplation that one can easily - very easily in fact - come to understand the many control factors in place that restrict the actions of 'humans in Universe' - and in a way funnel the activities of the modern wo/man into a smaller set of permissible actions than would otherwise be available to a free wo/man. This is the crisis Of now that I speak of; in that those

---

[102] the visual field of humans (and you k-now who you are) takes in about 10 frames per second from the visible light spectrum (and possibly from other parts of the electromagnetic spectrum in more subtle ways than an apparent picture show of the Objective reality that is in 'front' of us). Hertz is a unit for cycles per second, thus we can say our visual field operates at 10 Hertz.

sOb's make every effort to hijack the normative contemplation of humans (and you k-now who you are) and insist that the attention be paid to 'this' or 'that' (of the sOb's choosing), and we need pay attention nOw

thalamus: is this why so much attention in the media(s) was given to the comings and goings of the celebrities of our recent history?

Giordano: of course, this now makes total sense my dear herb; for as anyone can see, the comings and goings of the celebrities should matter not, be we humans (and you k-now who you are) are 'told' to pay attention to 'this,' or 'that,' and thus if we are focusing on 'this' or 'that,' we are also discouraged from paying attention to the sOb's behind the curtain[103].

Glia: and this seems to be just like the standard magicians trick of diverting attention from the real slight of hand.

Joe: right.

war Is peace

the next morning the interlocutors are quite refreshed, and many are frolicking about picking berries, fishing for trout, or foraging for mushrooms (no, not those mushrooms). After the morning meal, Giordano wants to continue with the crisis Of now, as he feels this is very related to the contemplation concerning 'on What is'

Giordano: my dear herb, the crisis Of now is a very real technique used by the cabal of maniac magicians, those sons of Beliel (the sOb's), to hijack the mental landscape of a vast majority of humans (and you k-now who you are). Do you have other specifics involving these type of techniques employed by the sOb's?

herb: the great shaman Orwell famously presented the idea of 'doublespeak' in his wonderous freeedom enchantment '1984.'

---

[103] because that would be the ever evil conspiracy theory thinking, and "everyone k-nows there is no such thing as conspiracy" - mumbles (the goddess) Sophia.

For example, if one is allowed to say 'war Is peace,' or to say 'freedom Is slavery,' then we have to agree at that point that words have no meaning whatsoever.

Giordano: 'war' is a word that can be thought of as the opposite of 'peace,' so if one says 'war Is peace;' would not everyone just fall over laughing, because obviously it is absurd to say such a thing.

AlBe: and yet it was the case, in our recent history, that that hegemonic country of herb's birth typically brought democracy with bombs

herb: thank you my dear AlBe, and exactly my dear Giordano, this is what a thinking person would expect. However, this idea that I have just introduced is an example of 'cognitive dissonance,' which is a tool employed by the sOb's to really damage the thinking process of humans (and you k-now who you are).

Because, as you imply my dear Giordano, if somehow two opposite ideas can be said to be the same idea, then sadly thought would have no meaning.

Giordano: and thus, if the sOb's can convince humans (and you k-now who you are) that thought had no meaning, then humans (and you k-now who you are) would simply stop thinking.

thalamus: and lo and behold, the typical humans (and you k-now who you are) divine right to contemplate the transcendental and the divine is less likely to be exercised ...

herb: because, via cognitive dissonance, the sOb's could have convinced us humans (and you k-now who you are) not to contemplate at all.

Giordano: because contemplation is not desirable when 'ignorance Is strength.'

wag The dog

> *Giordano is very excited to understand the depths to which the sons of Beliel will go to cast a net over the mental world of humans, and attempt to separate them from the divine... cue the goddess*

Sophia: another area, besides the technique of cognitive dissonance, that is effectively employed by the sons of Beliel (those sOb's), is the anti-human principle k-nown as the 'Hegelian Dialectic.'

thalamus: so, instead of a dog wagging it's tail, the tail here is in fact effectively wagging The dog?

Sophia: indeed my dear thalamus. The dialectic approach to the implementation of (more) control would first 'create' a problem. Typically, this is done using 'agent provocateurs,' or a full-fledged 'false Flag' attack.

herb: ahh, the infamous 'false Flag' operation, where the instigators wear the colors of another tribe in order to hoist blame onto that tribe.

Sophia: right my dear herb. So, a problem is created so that humans (and you k-now who you are), by design, would rise up and demand a solution to this new problem that was, again, do not forget, designed exactly to generate a fear and/or panic into the general population of humans (and you k-now who you are)

Glia: and why would the sons of Beliel (those sOb's) go to all this trouble to create a new problem my dear (the goddess) Sophia? are not there enough 'problems' already?

Sophia: because, my dear Glia, those sOb's already had the solution to the problem in mind when they (the sOb's) created the problem to begin with.

herb: and this, if you do not mind (the goddess) Sophia, is the classic problem - reaction - solution dialectic that is used all to often by the sOb's to bring in new forms of control onto us humans (and you k-now who you are) with the ever more ridiculous solutions to problems that are typically not real problems to begin with.

AlBe: and this happened when the sOb's, through their media minions, starting making a big deal of 'fake news.'

Giordano: and then the solution to this problem of 'fake news' was a campaign of censorship of almost any 'news' outside of the media minions' control

herb: even though it was the media minions who in fact were the main purveyors of 'fake news' (and propaganda)
Sophia: problem - reaction - solution in order to get more control
Joe: right.

the Order of the day

> *it is nearing lunch time for the interlocutors and herb takes a smoke break, and contemplates how much himself and others are still susceptible to the archonic locomotive breath, and associated techniques, even when they k-now those techniques are being deployed against them (by the sons of Beliel, their agent provocateurs, and/or various other minions - many of which k-now not what they do, or who they really serve). The discussion continues without herb, while some of the interlocutors prepare a midday meal*

Fate: when the sons of Beliel (those sOb's) are not working their Hegelian Dialectic to implement yet more control mechanisms, or practicing cognitive dissonance techniques in an attempt to persuade humans (and you k-now who you are) to just stop thinking altogether, 'the Order of the day' is then, typically, just to simply 'hate Each other.'

For you see, if the sOb's can fan the flames of hatred, and convince humans (and you k-now who you are) to hate Each other (for this or that reason), then those same humans (and you k-now who you are) will not be paying attention to the sOb's 'behind the curtain' (so to say).

AlBe: is this what the short-lived phenomena of the 'social Justice warrior' (sJw) was all about?

Fate: indeed my dear AlBe. The sJw instances typically involved paid shrills (or agent provocateurs) to organize thinking impaired individuals to fan the flames of hatred against any and all who had an opinion that might indicate any type of independent thinking.

thalamus: a definite crisis Of now technique it would seem.

Fate: exactly my dear thalamus; and if we step back a bit however, we can maybe sympathize with the sJw foot soldier, who, wanting deep down to help others[104], are instead tricked via a type of sOb mind-control to focus their angst against enemies of the sOb's. The sJw will thus direct a righteous angst not at sOb's themselves, or for example, direct that righteous angst towards the bankster minions of the sOb's

Giordano: and those bankster minions, through use of demonically inspired usury, are always pulling more from the public coffers as part of what the great shaman Bucky called the GRUNCH[105]

AlBe: and this 'Order of the day' for humans (and you k-now who you are) to hate Each other was accomplished in all kinds of areas it seems, and at all levels of society. For example, is this not also the story of the typical political structures that were found all about the globe?

Giordano: excellent observation my dear AlBe; so that instead of humans (and you k-now who you are) directing their angst at the banksters, for example, especially those banksters associated with the criminal organization called, don't you k-now, the 'Federal Reserve,'

thalamus: stealing 1 out of every 5 dollars for usury payments

Giordano: right, butt instead of this deception, started for the last time, as we mentioned, in 1913, humans (and you k-now who you are) were subjected to so much propaganda so that the majority felt obligated to choose one of two parties: either a Demoncrat, or a Republind

thalamus: as if there was any difference between these 'political parties,' both just playing their parts in the 'kabuki' theater production put on by the sOb's

AlBe: butt it was just another 'Order of the day,' a way to pit human (and you k-now who you are) against human (and you k-now who you are)

---

[103] and thus play the part of the archetypal 'catcher in the rye'
[105] from the shaman's last book, and it appropriately stands for GRand UNiversal Cash Heist

Sophia: so that humans (and you k-now who you are) would be trapping themselves in the hear and nOw, concerned with petty things, and thus a lot less likely to contemplate or even believe in things like an immortal soul.

old Time religion

herb has returned from a walk about just in time for the midday meal. After the meal, a hike back into the forest of our permitted ignorance is on order, as the interlocutors search again for the old ruins they had frolicked within not just a few moons before

herb: the great shaman Goethe is credited with saying

> None are more hopelessly enslaved than those who falsely believe they are free.
> - Goethe

And Goethe was a German philosopher extraordinaire, who wrote, among other things, an updated version of Marlowe's Faust.

Giordano: was not dr. Faust the character that 'sold his soul' to the Devil in return for the acquisition of k-nowledge?

herb: indeed, and in the Faustus mythos, the devil's name is 'Mephistopheles.'

thalamus: one of the many archons I suppose, or was Mephistopheles the demiurge himself? (Though 'him' is probably not the correct term eh?)

herb: a question for Mephistopheles 'himself' maybe?

Sophia: right.

herb: but Goethe did bring forward that 'old Time religion,' as Marlowe had done previous. As that practice of humans (and you k-now who you are) striking bargains with entities from the archonic realms is a form of that 'old Time religion.'

Glia: is that similar to praying to a / the god(s)?

Fate: quite similar indeed my dear Glia, in that prayers usually do sometimes 'ask for something.' Then, implicit in the asking is the accompanying pious devotion to the god(s) by the one doing the praying.

thalamus: eye thought the term 'old Time religion' referred to the fact that everyone was always blaming every problem on the Jew.

herb: yes, that my dear thalamus was one use of the term. Butt the conspiracy of an 'old Time religion' includes much much more than this; in fact, it includes many of the issues we have discussed previously in our dialogues concerning 'on What is.'

Giordano: should not each be free to choose how they want to think about the question 'on What is,' without following these or those rules, or simply copying how it was done before?

herb: and this would presume access to real k-nowledge of the mysteries

Fate: which the sons of Beliel (those sOb's) always insist are not for the profane

Giordano: or what about the stories of the magicians last words, that last piece of k-nowledge that they must past on?

herb: if each individual who would accept the most Dangerous of all undertakings were to have access

Sophia: then rightly, my dear herb. they could make up their own 'Mind' 'on What is'

Joe: right.

AlBe: does each individual have to pray to a / the god(s)? Is an individual human (and you k-now who you are) obliged to be thankful everyday for having a life in the / this physical world? and what about the doctor, was his deal with Mephistopheles good or bad?

herb: and it can be quite confusing my dear AlBe, butt like (the goddess) Sophia intimated, it would seem reasonable that each individual that is 'born into this world' have a choice on how to behave, and what they believe, and to whom they pray, or with whom they make deals

thalamus: and what about that 'good or bad' part of the query?

herb: and of course, without saying, at this point in time, each individual is apparently also free to choose theurgy or sorcery

Giordano: and making deals with Mephistopheles is surely sorcery

thalamus: of the most wicked kind

Giordano: witch would make a member of the cabal of maniac magicians proud

herb: and be that as it may, as each individual experiences their plight, hopefully with access to real information, and not the 'fake news' shoveled out of the tv by the sOb's; most, we can all agree, with reverence to their immortal soul, would choose theurgy over sorcery if they k-new magick was real

a Bit too much to think

> *the interlocutors find a somewhat hidden stairwell, covered with spiral and other types of geometrical markings. The stairwell disappears into what may be a cave. The interlocutors all feel that this could lead them to ruins*

herb: so while communication with the archons / god(s) / astral entities is indeed as old as human (and you k-now who you are) existence itself, as is the associated praying, and / or 'deal making,' we should be aware that the reciprocal control[106] attempted by these same entities is also that old

Sophia: and of pretty much the same nature as it is even in this day, and in the somewhat recent past of which we converse about so often it seems.

herb: and one of those control mechanisms of the 'old Time religion,' practiced by the sons of Beliel (those sOb's) also mind you, is the technique of 'social norming'

thalamus: ah, the 'Thought Police' [107]

---

[106] ... as a wise old one once was heard to say, 'when you stare into the abyss, sometimes the abyss stares back'

[107] fra 'Looks like you've had 'a Bit too much to think' & with a picture of a sOb in uniform, with sunglasses, don't you k-now

Fate: 'closed minds do stop thought crimes.'
Sophia: 'don't speak out or question.'
Giordano: and it should go without saying that it should be illegal or shameful to say or think things outside of the 'official narratives' given down to us by all the concerned officials who do their best to look after us
thalamus: and so when laws were passed, that certain things could not be discussed
AlBe: this had to be a good thing
Giordano: all in an effort of looking out for us
herb: lest our feelings get hurt when a truth is told
Sophia: 'ignorance Is strength' my dear herb, quit trying to be so bold

> *the interlocutors really yuk it up after this little fun, and a few - we will not say who - actually fall off the stairwell onto a floor of a large ...*

herb: what is this place, if not a meeting house of old

worthless Eaters

> *(the goddess) Sophia and (the goddess) Fate look to each other, and wonder whether the other is to tell more, as it concerns the Atlantians of plato, and / or their heirs ...*

Giordano: so my dear herb, are all these techniques employed by the sons of Beliel (those sOb's) just for kicks, or is there some 'end game' to the deceit?
herb: my dear Giordano, consider first what the shaman Huxley had said:

> The propagandist's purpose is to make one set of people forget that certain other sets of people are human.
> - Huxley

As the controlling of the goyim, us 'worthless eaters' (one should read that as to mean us peasants), requires much work and deception according to the Protocols of the Elders of Zion[108]

thalamus: if I may my dear herb, it is only after this type of propagandist's control has been implemented, can the humans (and you k-now who you are), specifically us peasants, be convinced to participate in war, where we kill other humans (and you k-now who you are) / peasants who have done nothing to us.

herb: 'always by way of deception'[109] my dear thalamus. And indeed, this ritual sacrifice of war, it is the modern way of how the sOb's make sacrifices to their controlling archon(s)

thalamus: no longer on a temple alter then, but by way of deception on the many battlefields orchestrated into existence by the sOb's?

Fate: and plans for war, the sOb's do make these plans centuries in advance

Sophia: and the banksters, with their demonic usury, make sure it happens

Giordano: all part of a Hegelian dialectic, and most likely involving a false Flag attack

herb: that 'old Time religion,' and 'the Order of the day'

AlBe: all because of propaganda and some weird mind-control, that we forget that others are human (and you k-now who you are), and then some are convinced to kill, which is a sacrifice of sorts

herb: and now the priest no longer delivers the new baby to the monster sitting on the the big thrown... butt they instead bomb a hospital or school

Sophia: and the archons will always still want more

---

[108] which may or may not be an authentic 'how-to' manual for the sons of Beliel (those sOb's), that cabal of maniac magicians, formed, mind you, from talmudic principles, right out of Babylon

[109] a war cry of one particular nasty bunch.

babylonian Money magick

the interlocutors continue investigating the subterranean ruins, witnessing geometric sigils everywhere, much like Egyptian ornaments in their temples of old. Joe starts a fire, in a fire pit, near what appears to be an alter of gold

herb: let's now go into more detail, of a very important, if not the most important tool, used almost exclusively by the sons of Beliel (those sOb's).

thalamus: this has to be related to something the banksters do, am I not right my dear herb? As it does seem that the banksters (servants of the sOb's all) do have lots of control.

Sophia: the babylonian Money magick technique of usury[110] is the form of control, brought out from 'the fall' of Babylon, is the old Time religion tool which hands the sOb's a type of demonic control

herb: and with total disregards for the 'Thought Police,' we should go full wyrd if we are to discuss this usury - a bane of human (and you k-now who you are) existence, we most certainly all would agree

Sophia: throughout history, usury has been against 'the law' in many places. Islam, for example, explicitly forbids participating in either side of an usurious transaction.

herb: the practice of usury (it can be presumed) was partly to blame for Jews being expelled from over 100 countries since, say, 250 AD.

thalamus: some of those countries, for example, France, many times over and over again in a short time span.

Giordano: the great Demon Cromwell rescinded the prohibition of Jews in England a short time before he also allowed a private Bank of England to re-boot.

herb: and it was the usurious sOb's who funded and backed the demonic Bolsheviks who wiped from this planet so many incarnated immortal souls

---

[110] def'n: the lending of money with an interest charge for its use; especially, the lending of money at exorbitant interest rates.

Joe: my dear herb, we all hear you denounce the banksters, and their usurious techniques, butt is there any other way for a bank to operate; I mean, can it operate without usury?

herb: very well poised my dear Joe. If in fact trade is not to revert only to barter relationships, we could in fact have a 'money system' setup which does not utilize usury.

Giordano: and oh how this would help stop war sacrifice, and many a planned genocide

> *(the goddess) Sophia is worried, if only because he wanted to print his own money, that born again shaman JFK, this is the reason why they killed him, can there be any doubts. And why were Syria, Iran, and Libya all attacked, because they would not accept a private central bank, and these are the facts*

Sophia: my dear herb, would you really risk life and limb, and dare mention 'a solution' to the sOb's old Time religion of (demonic / demiurgic / talmudic) usury?

herb: my dear (the goddess) Sophia, wyrd times require, do they not, wryd honesty to the predicament that humans (and you k-now who you are) find themselves in.

Sophia: leave it to the Fates then?

Fate: and if this be a destiny, you should 'make it so' my dear herb.

herb: thus, without any further ado, my dear Joe, we can nOw all consider how the money problem is solved.

We would require good record keeping within a group, this would be done by documenting, over selected time periods, the value of all goods produced – by the group, to produce, a 'group National product' (gNp), if you will. Then, an amount of money equal to this gNp is printed (over a selected time period), and used / distributed for the common needs of the group. This new money printed would not cause an inflation, or either a devaluation of the money in circulation, for this gNp printing simply represents, the fact that this much value in goods (produced by wo/man, or through divine mechanisms like plant growth – cough cough) has joined the group over the chosen time interval - let's say a year.

Joe: so we would want good record keeping, on the value of new goods / services produced, in order to make the gNp amount as large as we could

herb: here maybe others can add something, maybe something about an honest anarchy, butt the facts are that when goods are produced, some value is added, and this contributes to the gNp, and the next year more of own money is printed, because that represents the new value of goods that has been produced, with no need for usury (which bTw is usually charged on monies that the banksters do not really have)

thalamus: and who would control the distribution of these new monies, printed they would be without the need for usury?

AlBe: we would need some administrators for this type of honest anarchy

Glia: maybe an honest wo/man would not mind serving for a time in such a capacity

Giordano: and help watch out for the groups common supplies, like clean air and clean water

herb: oh my, there is a possibility that it could be so ($ 100,000) grand

AlBe: and if money can be printed this way, without the talamudic / demonic usury, then why is this not done?

herb: because this would take away a key technique of sOb control

Giordano: and the sOb's bomb their way to the insistence on a central bank

AlBe: 'money For nothing' and then the control is free, eye sea, and it is all so sad... butt why does not everyone k-now this state of affairs, and the chains that demonic usury places on wo/man?

psychology, Propaganda, and written history

Joe has the firepit going, and herb asks if the interlocutors might enjoy, a snack or a smoke, and then smiles as AlBe informs that she has found the remaining beer

Giordano: as it is with usury; psychology, propaganda, and written history are also a part of the Arsenal of very sharp tools available to the cabal of maniac magicians.

AlBe: you mean those sons of Beliel (the sOb's)?

Joe: as herb has explained it to me, the 'art' of psychology can involve serious 'black magic' / sorcery techniques, and combined with propaganda can many times get humans (and you k-now who you are) to do things they would otherwise not do (and definitely not 'want' to do).

Glia: 'don't don't don't mess with their will' is probably not a mantra they (those sOb's) sing about their campfires!

thalamus: it used to be illegal, one should k-now, for the government of the hegemonic country of herb's birth to run propaganda against it's own citizens.

AlBe: butt they, the sOb's, actually changed 'the law' to say it was now OK, to run propaganda against it's own citizens, eye think it was done by the military, or the CIA

thalamus: which of course they (those sOb's) had been doing for a very long time, and in that age of television (or tv), the sOb's had a very appropriately named 'Operation Mockingbird,' that facilitated the intelligence (sic) agencies control, of most of the media, and the (fake) 'news.'

Giordano: written history, also subject to Operation Mockingbird type influence, has always been an effective control mechanism, has it not my dear herb?

herb: all part of that old Time religion my dear Giordano, we should all so easily agree. And what of ULTRA, or OFTEN, it seems all so absurd. And what of the NATO officers captured in Aleppo during that war, hardly anyone was allowed to hear, the sOb's did not want you to k-now

Sophia: butt before we leave this topic of psychology, Propaganda, and written history, can we all ack-nowledge that this is the same as what lately we have been talking about

Giordano: indeed my dear (the goddess) Sophia, the only difference nOw is that we have another set of names for those efforts to implant, say, cognitive dissonance, and we can plainly see, that to hide oral traditions, and real history, we simply burn the old books (or hide away the real libraries), and provide a new written history, and then convince everyone, it is true, it is true what is written, 'can not you read man?'

Fate: what would you do with the dissenters my dear Giordano, who would not submit, two a new history that was written that was obviously false?

AlBe: the sOb's would probably exterminate them as a sacrifice, or lock them all up

Sophia: ah, re-education camps, that could not be fun

Glia: censorship, censorship, the courts and the law, make it illegal to challenge the 'news,' any school taught 'history,' or anything a government official says, does not that seem a good plan?

herb: oh how sad it all seems, deja vu, eye sea it happening all again ...

Now who is up for some adult beverages, a large helping of internet censorship to help wash it all down (down down) (dododo), we should investigate further this place we have in our destinies found

lilly Waves and mind hacking

> *it has been a few sun 'sets' since the interlocutors last discussed some of that old Time religion. They have all happened upon a particular radiant pool deep in the underground ruins that some of them are considering never leaving. The fauna and fruits are vast, as the interlocutors can not butt smile at (the goddess) Sophia and give thanks to (the goddess) Fate*

herb: another set of techniques that became standard issue in very ancient times, and in the not so ancient times, were techniques which would alter brain operations of individuals without them being aware of the alterations.

Giordano: right, and as I do recall, the shaman John Lilly identified certain types of electromagnetic techniques that were actually effective in bypassing frontal consciousness of humans (and you k-now who you are).

thalamus: and the many magnetic experiments done to try to develop straight-up black magick techniques for mind hacking.

Joe: and there were many other entrainment techniques developed to make influence on the mood and mind of wo/men without them being aware of it.

AlBe: i am hungry.

Glia: i am thristy.

herb: i am a peanut butter sandwich.

Sophia: butt even with all this nefarious influence, one big puzzle that is still left for humans (and you k-now who you are) to resolve is the fact that signaling from brain to body takes too long...

herb: so maybe my dear (the goddess) Sophia it is the case that the effects resulting from the techniques employed by the sons of Beliel (those sOb's) do not matter at all, because we ourselves only in the end have an epiphemiral existence[111]

Fate: we have discussed that issue of brain / body 'signals'

---

[111] thus, our 'reality' is just a motion picture show constructed especially for us individually

(whence 'Mind' interacts from Negative Universe) earlier as you well k-now my dear herb, and altogether have agreed that it is most likely the case that humans (and you k-now who you are) get backwards in time signaling similar in style, to that proposed by the shamen Penrose and the shaman Hameroff, via microtubules as they supposed

herb: right, and that saves us, and allows us to contemplate the need for an Objective reality.

Giordano: back almost to where we started then?

Fate: far from it my dear Giordano, because now through our mental gymnastics we have come to the realization that humans (and you k-now who you are) are also susceptible to many different types of entrainment techniques wherein the vast amounts of normal signaling that coordinates an individuals situation within an Objective reality can be tampered with.

Glia: for shame that someone would mess with the divine in such a matter.

Fate: for shame indeed my dear Glia.

Sophia: and as herb alluded to earlier my dear (the goddess) Fate, these entrainment type techniques are extremely ancient.

For I k-now of a time when the ancient temples surrounding the Mediterranean sea were used as a amplifier (of sorts) that would send out both audible and inaudible sounds that had a very specific set of influences on those in service of the particular god(s) of those regions. So you see do you not; they k-new magick and sorcery in those ancient of days.

thalamus: the same cabal of maniac magicians, our foes the sOb's?

Sophia: we should maybe let the Fates decide that my dear thalamus.

Giordano: and what was accomplished with the temples in the ancient of days is lately, I suppose, being accomplished with radio, television, and many many hidden signals that we are immersed in

herb: butt some of these signals used to entrain individuals are truly not hidden at all. Recall, some of you, how it was that the television 'shows,' and 'movies' of recent times would regularly show scenes of such vulgar violence and bloodshed as to make people numb; numb not just to such unneeded[112] horrific violence, butt numb to feeling in general. This numbness came partially about due to the overwhelming 'fake reality' hoisted upon each individual, and thus their individual reality-tunnels, without proper training, would adjust just in order for each human (and you k-now who you are) not to make themselves sick.

Giordano: and that is a form of entrainment; which is so clear to me now

herb: and the 'shows' that projected the many fabled 'law enforcement' groups who – oddly enough to the thinking wo/man – 'broke the law' on a regular basis in order to enforce the law... as a matter of course.

Sophia: 'war Is peace,' 'freedom Is slavery,' is very close to this idea that 'only law makers and law enforcers can break the law,' eh?

thalamus: thus entraining humans (and you k-now who you are) to simply submit to the sOb's and their farcical 'law.'

AlBe: the 'law' is the chains that bind you my dear thalamus, and if you have any doubt, think of all the innocents locked up over the years for smoking a plant

herb: and this is a very important and subtle point that our dear AlBe makes, for you sea, can you not, that for a general population of humans (and you k-now who you are) to accept a tyranny that restricts access, according to the (farcical) 'law,' to a plant that you can grow in your garden, this has to be one of the most ridiculous things of all!

Glia: so is this where we squat down on our knees, and then prostrate ourselves to these sOb's, because their sorcery must indeed be very powerful, if they (those sOb's), can get away with restricting access to this plant, that has many divine aspects when processed by you or me?

herb: a capitulation then, has it got to that point, where there is no turning back, and we admit theurgy has lost the fight?

---

[112] and anti-human such violence is... further evidence on the 'hint parade' as to how the sons of Beliel (those sOb's) still had everyone 'by The balls'

Sophia: what fight is that my dear herb, just the ability to 'smoke pot'? or are you concerned with the bigger picture, say, for example, your individual struggle to be an archetypal 'catcher in the rye'?

Giordano: eye can not speak for my dear herb, butt I 'will' give it a try, and say it is more than just about the 'law' (those chains that bind), and in the end it may (or June) be about each individual's immortal soul

on Mind control and Henry Ford's assertion that 'all history is bunk'

> *the underground ruins keep providing surprises for the interlocutors. After a band played a few interesting songs, in a wonderful amphitheater, they attempt to wrap up the discussion, concerning 'on What is'*

herb: a shaman Jack Heart opined that he agreed with the shaman Jung that

> Time has no power over the spirit and will of man.
> - C.G. Jung

Giordano: but the shaman Denis Healy, a former 'Minister' of Defense for a stronghold of the sons of Beliel (those sOb's) was quite serious when he stated

> World events are staged and managed by those who hold the purse strings.
> - Denis Healy

thalamus: and this is probably why the shaman Henry Ford was fond of saying

> All history is bunk.

herb: what then is one to do when confronted with 'on What is?' Do we simply submit to mind control, and try to live out our life on / in the / this physical realm in as much luxury as possible? or do we confront the sOb's and their excessive mind control efforts with defiance, and attempt to think for ourselves?

Giordano: butt how does one go about thinking for themselves my dear herb, when everywhere and everywhen is disinformation designed - it seems - to 'throw us off the scent' of 'on What is'?

Joe: my dear Giordano, have not our many discussions helped you realize that a dedicated individual, or group, if resigned to this most Dangerous of all undertakings, can free themselves from the mind control and start to think deeply concerning 'on What is,' probably with a little help from real friends[113]

herb: correlate my dear Giordano, correlate. Anyone who tries can plainly see, all the evidence is in front of them, because the sOb's do not feel the need to hide it anymore. Butt of course one must be brave. Here, for example, is a snippet, a small freedom enchantment if you 'will,' from the space-shaman Gorden (of runesoup(.com)) which describes the situation quite well

> I'm not really interested in discussing whether these things are happening or not - particularly when it comes to narrative manipulation - because that ship has well and truly sailed. They are happening. It takes bravery to face that. I'm more interested in exploring how others correlate these actualities in their mind. It seems to me we exist in a universe that is massively shit, massively amazing and whose macro story is probably so incredible it could melt your face with wonder.
>
> Never stop correlating.
> - Gorden White

---

[113] and / or a freedom enchantment or 3

And it is the correlating of all the evidence; some hidden, butt much of it not So hidden, that will allow anyone to ponder, correctly, and sanely, on the question of 'on What is'

Glia: and would it not help tremendously my dear herb if many of the seckret society folks would come clean with the information they keep in their groups, 'on What is,' hidden history, and the whole damn lot

herb: indeed my dear Glia, and very germane to this is an important bit of advice we should give anyone interested in embarking on an investigation into 'on What is'

thalamus: what, 'never go to a party where you can not dance'?

herb: no my dear thalamus, even though that may be the best advice anyone could ever give anyone! No, here I would remind everyone of the advice we heard earlier, and which can give you a free pass (in a similar way to how the poor widow's son carries on, I can only suppose). K-now nOw that this advice is useful when one is feeling pressure to join a group, witch keep sekrets, one is allowed to always say

> I don't know if I want to be a member of a club that would have me as a member.
> - Groucho Marx (attributed to)

Giordano: so you could not join a military group, a corporate group, a masonic group, a local coven, or any of these types?

herb: so does that tell us something about who does, and who does not serve the sOb's; for if it is seckrets they keep from the profane, us peasants, then eye would suggest you not give them a second thought

a Hidden religion?

near the amphitheater where the band has finished playing that wonderful freedom enchantment 'Peace, Love, and Understanding,' the interlocutors find a gleaming walkway weaving its way towards a large temple of sorts. When the interlocutors enter, it seems

> *spirits of some kind materialize throughout (like in the ancient Mulan cartoon); but it creates no fear, as it seems intended to do, the interlocutors are all well versed, in a helpful mantra, that the interlocutors repeat three times: 'no ego, no envy, no fear'*

AlBe: what are these 'creatures?'

Sophia: why not ask 'them' my dear AlBe; as it is likely you may have met at least one of them, maybe in your dreams, before

AlBe: what are 'you?'

> *'we' are a species of astral entities who would normally have great difficulty materializing in your (physical) realm*

thalamus: my dear herb, are these the archons you have spoken of before?

herb: quite possible my dear thalamus, butt I have to wonder how it is that so many are here nOw, visible in a waking, non-sleeping, and non-meditative state.

Fate: this temple is a special place of refuge for 'many' from other realms

Giordano: built by 'whom,' or by 'what?'

herb: most likely some type of Atlantean sorcery from the ancient of days my dear Giordano

Sophia: and that is it my dear herb, butt a type of sorcery that has indeed lasted down the ages, and as the Fates would have it, 'they' are here for the interlocutors, nOw

> *if it pleases the interlocutors, 'we' would welcome the chance to inform you and tell it like it is, and note that 'we' stay involved in the unfolding history of humans, and as herb constantly reminds 'us' all, you k-now who you are*

herb: it would please us, and do tell more, about how the sons of Beliel (those sOb's) work with archons as part of many a Hidden religion ritual

Fate: and this will be done without deception and lies, as 'you' k-now very well, that this is not a 'time,' or 'place,' for that

> *it is true that 'our' kind have been conversing (in a way) with the sOb's over many different aeons, and in many different forms. The Brahman of Ancient India, the Magi of Persia, the Kabbalists, and many others in the magical arts: both sorcerers and theurgists, k-now how to establish contact*

AlBe: and in what way then would herb be accurate in saying 'you' are involved, with the sOb's, in 'a Hidden religion,' of sorts?

> *because as it is, those sOb's who have had a hidden hand in all affairs involving humans (and you k-now who you are) for longer than your written records indicate, have gained guidance and help from 'some' of 'our' kind*

AlBe: for good or for bad?

> *and here is maybe where the problem is, as 'we' may not have these same type of conceptions as it concerns your realm*

herb: as I have intimated many a time, 'they' are all demons; those archons who would agree, to interfere in the goings on's of humans (and you k-now who you are), and thus assist in a sOb's sorcery

Sophia: and what of the archon 'Prometheus,' who brought fire, and other k-nowledge, to humans (and you k-now who you are) in the fabled ancient of days?

Fate: and what of the muses who have assisted seekers over the years, with 'hints' and 'maps' (of sorts) in their dreams and meditations?

herb: that 'they' would let 'themselves' be summoned to interact in this realm by techniques of Kabbalists / the Dark magus John Dee's Enochian magick / and other sorcery techniques lends credence to my assertion that they are part of the sorcery

> *and guilty 'we' are of playing these games with representatives of the sons of Beliel; butt, as (the goddess) Sophia has suggested, 'we' are also guilty of providing useful information to humans (and you k-now who you are) in ways that would not be considered sorcery, butt as a statue or two has indicated to you my dear herb, we are also willing to help with theurgy*

Sophia: are there any hints 'you' could provide my dear herb, that would assist him and the interlocutors, in their quest concerning 'on What is'?

Fate: and do tell us the mechanisms by which humans (and you k-now who you are) could put a stop to 'your' constant meddling in their affairs

> *while 'desire' and 'purpose' are not in 'our' nature, 'we' can tell the interlocutors that not all of 'us' are pleased with the part 'we' play in the affairs of humans (and you k-now who you are).*
>
> *It is often the case you should k-now that when an immortal soul leaves their incarnation in the physical realm that 'we' will answer queries when under interrogation by these aforementioned immortal souls. Butt not always mind you.*
>
> *Not all of 'us' understand the hierarchies of the realms, and unlike humans (and you k-now who you are), many of 'us' do not seem capable of doing anything other than what 'we' are told*

herb: butt all this too may come to past, as it may be the case that when it becomes general available k-nowledge as it concerns the nature of this particular archonic realm, that humans (and you k-now who you are) may not be so susceptible to the incessant 'archonic breath.'

Sophia: it was the case in the ancient of days, that many a theurgist attempted to imprison these creatures, and block 'their' access to the 'Minds' of wo/man

Fate: butt this has always, always been unsuccessful, since, as herb intuited, we never came to a point of global education as it concerns the immortality of the soul, and the ontological connectiveness that humans (and you k-now who you are) enjoy with all the realms, as a matter of course

herb: and we should not be surprised then if the Great Demon Jehovah has much to do with maintaining this deception, and these lies. Even in recent times, the Demon and the sons of Beliel (those sOb's), continue to teach all humans (and you k-now who you are) that only the material Universe is real; except of course beyond their insistence of the 'reality' of this one particular god / the Demon 'himself,' who, don't you k-now, maintains at least one yearly festival that celebrates genocide

Giordano: lies, lies, lies, that seem to be allowed, and / or encouraged by the sOb's, and all the minions 'they' have on their side

> *it is their real religion, and to you my dear herb, 'we' are sorry that we can not address this typically insightful observation, in more detail*

Sophia: butt we should not despair my dear herb, butt rather, continue the defiance that truly is the only answer to the condition of the absurd that the interlocutors have managed, in their religious documents, to describe so well

Fate: and be advised, that as our dear herb intimated, these rules 'you' have, these things and influences that are done on behalf of the sOb's, these too will pass as time flows from the future at the same 'time' time flows from the past

> *may it please the interlocutors to recall that not 'all' in our realms would choose to side with the sons of*

> *Beliel (those that you call the sOb's). There are some of 'us' that find the effects of sorcery on humans (and you k-now who you are) quite distasteful, and yes, absurd*

herb: and what is to be done, or what has been done, that would assist us humans (and you k-now who you are) against the sorcery of the sOb's, a Hidden religion, 'their' sacrifices and genocide

> *your temporal ambiguity is well suited to such a query; since, as you see my dear herb, the temporal sense 'cast upon' the typical wo/man may have its rationale deep in sorcery, in other words, as the great shaman Knowles intimated, humans (and you k-now who you are) are still under Saturn's spell. Many of 'us' do not understand the flow of human (and you k-now who you are) affairs in quite the same one-thing-after-another type way, but 'we' do k-now the forever 'nOw'*

herb: and this is of course the way it may have to be when the truth of the immortal soul is cast into the thinking. As the shaman Arthur Young, and many others, k-new, 'Mind' truly can k-now the past, present, and futures all at once. For how else could theurgy work, if not for the ability of 'sending' our intentions, via hooks, 'back in time'?

Sophia: notwithstanding the physics lesson my dear herb, we may be able to garner for the interlocutors more insightful information still

> *(the goddess) Sophia is correct, and we are herewith authorized to inform herb and the interlocutors that there are many, many 'things' at play in 'our' realms as it concerns your free will*

> *To wit: not only has Jove kicked Mars's and 'his' war machine of 'heaven' to the curb, but the chains that bound many a god(s)[114] have been loosened and / or been altogether removed*

herb: at last, it is as in accordance with the theurgists 'will', the divine can / does provide counsel to all 'things' along their course of destiny, in association with (the goddess) Necessity.

Giordano: about 'time' that 'heaven' is to be reordered.

Glia: butt, my dear Giordano and my dear herb, does this reordering of heaven require the typical phase of genocide, that seems in 'history' to be part of any world change? Is not banishment better, that would be my 'will'

thalamus: and what of the war sacrifice practiced by the sOb's, will these atrocities too come to pass?

> *my dear Glia, we pay honor to your query and concern, and we can say that some of 'us' would have it no other way. In situations that you have 'nOw,' which one of 'us' can say, what will be / has been / can be involved*

Sophia: 'you' tell us straight

Fate: and without deception and lies

Sophia: the answer to our dear thalamus and our dear Glia's query, about sacrifice and genocide

---

[114] certain god(s) in fact, that would support the humans (and you k-now who you are) and their desire of deliverance from the sorcery of the sons of Beliel

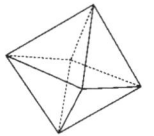

INTER-MISSION

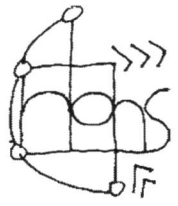

time Fades

covered by water, the ground, earth wearing and forming under the tremendous forces which exist above it and around it. gravity and events play their hand and the view for the future people begins to expose itself. feet begin to hit the ground as does the flow of water as what will be known as rivers, streams, lakes; and other names from other tongues. the distances are great and variety is also, the events continue as they did the prior sun day and moon night. the earth is rough and unforgiving, lifes abilities are tested time and time again as not much changes but the ideas of wo/man

time Fades

berries picked, shelters found, much learned; about the land and sky and sea, and the cycle of things surrounding them. groups of creatures on land, in the sea, in the air; not one talking to each other, but they can see, smell and hear; their neighbors, the groups of creatures on land, in the sea, in the air

times Fade (tFtFtF)

as was stated, time fades, christmas has passed, civilations come, most do not last. a poem, a story or just a feable tale; things to do and say as the times fade

a Tale of human existence in the land of shadows, or how wo/man saw light And life (A pOpol vuH)

Every wo/man is a star - (by way of Fra. P.)

the Mayan pOpol vuH, or what we had left of it, starts with a story of two brothers who more or less are intent on outsmarting their overlords, a bunch of predatory, self-centered, but omnipotent mind you, keepers of the realm. herb found the story somewhat familiar to other "religious texts" that could be found about in the world that he had been born into. Unlike other so-called religious texts, except maybe some of the Gnostic tomes (which herb somehow intuited were in fact insightful), these other religious tomes had a little too much angst, and as herb surmised, too much fear. So it was that the pOpol vuH had more a striving tone, and against the odds (the overwhelming odds it seemed), it was OK to strive, and it was OK to smile, and it was OK to love (ooo).

In the days after many a grand discussion on Physics, on Mind, and on What is, herb pondered many a moon about how best to explain it all over again. This should of course be done without resorting to hidden techniques of propaganda, which of course was sorcery, which all should eschew, and each should rather pick theurgy, the divine magic, for their doings, writings, and thoughts. So how was it then that herb could close out this freedom enchantment, when it was plain for all to sea, that the physical was not the only realm, and it was still quite mysterious, how we all came to be.

herb went back and back and back in time, in his mind, in those experiments, in those dreams, in those walks, and there came many a realizations, and many a smiling thought. herb understood that language was in the beginning a beautiful, magical or theurgetic "thing," the sounds, tones, and inclinations of a speaker, it brought forth much insight into the human (and you k-now who you are) condition and the individuals current plight. Word sequences, if one could call ancient utterances words, which we should not, but these sequences, they came forth from a connectiveness to "things," and maybe, just maybe, each utterance was indeed a religious or mystical "thought." How sad herb could become, if he ever allowed himself to be, when he stepped back into his time in that hegemonic country (of his birth), where he realized most "things" about ancient languages had been destroyed, and that "language" was now more or less only about control, and that doublespeak, Newspeak, whatever that was that the great shaman Orwell wrote, this was now the norm, and the divine, don't you k-now, weeps when anyone understands, the fact that modern "language" was now, intended, in a subtle way, in fact it, i.e., modern language, was designed that way, as many a Dark magus intended, our language was designed to kill "thought."

the pOpol vuH, as with all religious tomes, whether explicitly or not, holds as a basic tenet that there are, in all of the divine creation, more sentient beings than just humans (and you k-now who you are). Sometimes, in some tomes, there is only one other entity besides humans (and you k-now who you are), and in yet other tomes there are many more. Certain tales, for example, suggest the shaman Solomon (yes, the one of biblical fame) enchained 72 demons and tasked them with the building of a temple, and these demons you sea, they were not human (and you k-now who you are), and in the story (or tale), Solomon did indeed bring them forth from another realm. Stories of angels and demons, tribulations and waiting rooms, or coming before a judge, or weather one has a place in heaven and / or hell, these are all stories, mind you, witch in fact demand some other realm(s). So then the question, is it OK if we ask(?), what is a human (and you k-now who you are) to do when and if they accept such a thing, the thing is, we mean, that this physical realm, which is maybe the lowest of all, populated with books, people, and plants, and other "things" that some are sworn not to tell; what is a human (and you k-now who you are) to do, for a lack of a better phrase, as they walk in this "land of shadows" (witch is this physical realm)?

there is an old way of thinking about all that is, by some names it is called the "doctrine of emanations," and yes, it can help explain that apparently tricky business about your immortal soul. For herb's purposes, as he sought yet another way, to explain that freedom enchantment, that desire, which many have, one wood surmise, a way to explain to all wo/man that they are a star, and the emanations and cells and signals that each star has, is coordinated in ways that frankly no one should be able to understand, but it involves those K-nots of 'light', and not just in this realm, but all the way back to when it was all still.. an urzeit if u will

Other stories bring forth the physical Universe with a "word," and not a word in the current usage, but one that rang out, established a vibrational basis, one would have to suppose, a "word" that on top of which all "emanations" would play, yes, even those "emanations" that make up you and me. herb liked this idea, because it had an ontology, some basis, however tenuous, that is all you need, to develop an ontology of how it is all built. Akasa was also an ancient name, the stuff from which everything came, a Pleroma of the Gnostics, pagans and mystics, and also an aether to some. herb surmised whether it be a word, a Pleroma, or many layers of astral realms, there was some connections that allowed a human (and you k-now who you are) to connect to it all, both Forward and backward in time, or if one choose, these allowed connections could be used for the proverbial nOw. If each one and all 'things" used little K-nots of 'light,' to represent the divine emanations that, frankly, encompass each individuals plight, then what choice does one have when they ponder individually, how to continue correctly in this "land of shadows," or better yet, how can wo/man sea light And life?

as many k-now, and a few reported to herb, that they can see in their dreams, different faces, different places, sometimes it seems so real. Many ponder if they go somewhere, or do they stay in this physical realm, is the question really hard, or could a child understand if taught about it well. There are many souls who k-now deep down, that their current life, here, now, in this physical realm, is not the first or the last, meaning they can have many more, butt do they have a choice, that, many are not willing to tell.

when some ancient books were put together, it brought information and k-nowledge from here and far, one, the Kabbalah, out of Babylon, had many instructions and observations, butt let me tell you now, those writers and plagiarists, they would most likely delight, if some of that information they had put together had never saw the light, meaning they wanted to keep it all under wraps, those stories of how all the realms are attached. This information you see has two or many forms, some for the public, some for mystics, and other forms mainly for deceit.

so then in this "land of shadows" that is the physical realm, do we have an overlord, or a (set of) benevolent god(s)? is it possible that a demiurge has control of each and every thought? or are you individually an immortal soul? are we each part of a greater thing? or are we just meat, that when it quits functioning, will begin to smell? oh my oh my, herb had better write some more ...

those Divine emanations

the idea of Divine emanations matched up well with the shaman Arthur Young's little K-nots of 'light,' and those Divine emanations were not just in the physical, it was the basis for all things, future and past, as those Divine emanations also make up you and me. According to herb's reasoning then, those Divine emanations came to "things" through a reality flux, that connection between the physical realm and all that is 'above and below.'

"Every thing is as it should be," was a description of the magical reality of what is, so surmised the shaman Crowley, but it is really hard to understand. Here, this would seem to suggest, that all could be determined before it happens; so how can we understand it, that if I choose one of two things, that this is what was to be? who decides? or it is a decision not to accept responsibility; and what of the here and now (they scream at the top of their lungs)?

Butt something else you should k-now, since, indeed, you are a star

rules 4 theurgy

> *If it is not true, don't say it*
> *If it is not right, don't do it*
> *If it is not yours, don't take it*
> - the Cowboy way (by way of The Virginian)

herb had a friend whom he often referred to as "my friend the psychonaut," (mftp)[115] butt this was a long name to continue to always use, so herb made a sigil, and when red, the sigil was plain to sea, it said tEd. It was tEd who talked to herb about the "Cowboy way," and herb much enjoyed that first line for sure, to wit "if it is not true, don't say it." To say something that is not true, k-nowingly, surely that is a lie, and a lie is a type of sorcery, is not that easy for just about anyone to tell? Sorcery, in the sense, that the lie attempts to bend, it attempts to bend reality, from what is to a place or time where "What is" is no longer the case, but rather, the lie is what we are supposed to pretend.

so it was then also, that herb had come to understand, that for any interaction with the divine, or an interaction with the sons of Beliel (those sOb's), so it was that about any interaction at all could be turned to either theurgy or sorcery; depending, eh, solely on one's intent. If the intent is to control, surely that is sorcery, and that second line of the Cowboy way definitely did address that, for do not you sea, "if it is not right, don't do it" is a mantra for theurgy.

---

[115] borrowed from the space-shaman Gordon, who has his mmtp - "my mother the psychonaut"

herb would many times discuss with tEd, concerning the issue of divine counsel, you k-now, those hints and helps we have access to along our course of destiny, in association with necessity. Butt tEd did not like the word 'destiny', and tEd was not sure about that idea of an immortal soul, butt for sure, he was convinced, there was indeed more than just this physical realm.

if there is not an individual destiny set for every star, you k-now, you, dear reader, the current incarnation of your immortal soul; is there just your current plight, or is all divine counsel, and those Divine emanations that drive all, is it not a "mandate of heaven," a query of sorts along the line of "who will hold up the foundation of it all"? It may be left (or right) to each and every theurgist, possibly resident in every absurdist, eye am sure you understand, because is it not absurd to "think" you can effect things, that is, eye mean, effect the ways of the hole creation with each and every thought.

herb would ask tEd, what should he do, when allowing divine counsel, should he use it as a hint, or just do what he is told. tEd suggested we can each ask for help, we can ask for assistance to help us on our way, we can ask for assistance to be a better wo/man. With this herb concurred, for herb did not think it should be an interaction like the fable of old, where an interaction that started this way turned into a Faustian bargain, like the one Goethe wrote about, while elaborating on Marlowe's story, about, as it turns out, about a certain type of sorcery. That was the other choice then, was it not, that one could ask for divine counsel so that they can be assured of each and every thought; butt if this be the case, surely divine counsel it is not. Dr. Faust took this route, that is surely sorcery, he made a type of bargain with a demiurge Mephistopheles, and this sorcery allowed, Faust to visit a future, a here and now, and to have certainty.

"If it is not yours, don't take it" was the last line of the Cowboy way, and surely this was not sorcery, butt rather in line with theurgy, for if you follow a path in line with theurgy, you need not take things that you really do not want, or that you really do not need. tEd suggested to herb, cough cough, that he should just ask for strength, a strength to enable you to do the the right things, a strength to continue in a Cowboy way. Here then it is OK to question and challenge, and OK to hope and dream (OO). And most important of all, though we may strive to fill a role, of that archetypal 'catcher In the rye," whom would save those about to go over a cliff, we must always be on our guard, not to be also along the way always telling others how to behave.

back 2 ontology

The 'noise of the world' comes to each and everyone, if they allow a thought or two to mingle with the here and nOw. herb began his ontology talking about how a computer worked, that flow of that mysterious fluid type magical substance called electricity, which we reasoned was also a flow of number emanating with little K-nots of 'light.' The ancient Chaldaean Oracles, from a mystical past, teachings brought forth through the aeons from father to son, and in these teachings there was only a solar world where endless light subsists; so you see, with the Chaldaean system, the sacred reason may have an ontology. The physical realm, put together by a Demiurgus, the artificer of the Universe, allowed the immortal soul to be wrapped in a garment of divine light, and this patterning, we can only surmise, is handled via the reality flux.

when considering the reality of what makes up an individual's thoughts, we must, or should consider, whether we like it or not, those techniques the occultists use, and then we should also consider whether what they in fact do is theurgy or sorcery. Even the Quakers, and those who practice Eckankar, and many others you see, make attempts to visit and be absorbed by an astral light, as did the shaman William Blake, one simply lets it fall onto them. It could be theurgy, these type of techniques, when you seek divine counsel, while at the same time are not looking to do what one is told. Now, these techniques of immersion, in a kind of light, it can happen in meditation, and it can happen at night (butt may be best done in a Faraday cage, if you have one around). Now there are many other things an occultist can do when they set out to work in other realms, and sadly, at least to herb, some of what occurs, and has occurred, really is / was sorcery.

herb and the interlocutors had discussed the symbols and systems used by the Dark magus John Dee, a set of Enochian symbols, attributed to Enoch of old, and frankly, many supposed, it was a way to communicate with a divinity, or so we are told. Butt many secret society, who by the by, probably do not roll the Cowboy way, these groups have techniques and secrets, and many are not allowed to tell. And what is it you may rightly ask, that populates these other realms? The Gnostics called them Archons, oh my, herb reckoned them all demons, which is of course a word, like we have supposed before, a word to describe that witch could answer a spell. Chaldaean considerations included the higher demons, or "angels" above the human soul, and above that, as part of the 'empyrean,' there was the planetary dieties, the archangels and the Unzoned gods. Then, it was surmised, that for each human (and you k-now who you are), there were three 'souls'[116] that made up each and everyone: a divine soul that was immortal, a rational soul that could become so, and lastly an irrational soul (that maybe was from a wayback machine – a type of reptilian past, or maybe a hidden hook, that the sons Of Beliel – those sOb's – can use to manipulate human 'nature' so that they do not act like humans (and you k-now who you are) any more). Butt recall those daemons of Socrates, and in that lore we are not really sure, the nature of each type of influence that one can get from afar. Some

---

[116] many a Chaos magicians in the world at the time herb was resident in that hegemonic country of his birth, would say there are, say, 7 different influences operating in each and every one, and when you fashion a sigil, or perform a ritual, you may in fact be sending a signal to one of the Nine. The shaman Jung may not have agreed on an exact number, butt he had his archetypes, which, one can surmise, provided a scaffolding for such thoughts, especially those experiences in dreams and synchronicity. The great shaman William James, oft forgotten because he reasoned well that there was an immortal soul, what would he say about the sons of Beliel (those sOb's) obfuscation of our relation to these other realms? And still many other classifications of what is soul, what is spirit, and what stays or goes when an individual's time is up, all in an attempt to pinpoint, we can reckon, the divine part of each and every star.

strict materialist could insist that any such experience is due only to the subconsciousness, resulting only from events in physical Universe (or lowest realm), and whatever be the case, this is part of an ontology (if there can be one) of this physical realm, and of each and every wo/man.

this set of dealings with entities that are not physical, and that we can not smell, apparently has always been a part of theurgy, in that the theurgist ack-nowledges that other spirits exist, and many tales of old suggest some could animate a statue, like the ones the Greeks built, and there was even a famous play, that Sherlock Holmes surely loved, wherein it happened that way. herb was not sure of all that, the statues and all, butt herb himself would tell you that he did occasionally talk to rocks.[117] Are there rules of engagement, that is to say, can these other spirits be summoned, for example, with a magician's spell? oh my, and would this be theurgy or sorcery? In the parlance of many magicians both of old, and even ones today, there are many ways to reason about how the contacts should proceed; and can we all not sea, that this is part of any real ontology, when you accept that there is more than just the physical realm. When a magician speaks of evocation, their purpose simply stated is to bring that demon forth, yes, out from another realm. When some magicians speak of invocation, oh my, it seems much worse, since here the one calling is opening up, supposedly so that the entity (yes, from another realm) can take over their being, for how long who can tell. Invocation could lead to a type of 'possession,' witch is surely sorcery, some even call them 'walk-ins,' those entities that would be so bold, to take control of a human (and you k-now who you are) in such a way, that surely it would seem, the divine should not allow. Butt we do not make the rules, that seems easy to sea, and trust herb on this now, when he tells you that it can happen this way.

---

[117] and some told stories of coming from afar, from that great water planet called Tiamet, and when it smashed up in that famous (butt hidden) celestial battle of ancient times, these many river rocks just floated in space until arriving on this planet along with many frogs… or so 'they' say

So in this 'land of shadows' that is the physical realm, we should be aware of some of those techniques employed by the sOb's (yes, them, the sons of Beliel). There are many grimoures, books of magical 'spells,' with only a few, butt not many, of them freedom enchantments[118], most of them, one could conclude, contain a black magic of sorts, especially when the purpose is control. Butt consider something deeper dear reader, when we point out, the 'doctrine of emanations' may allow all sorts of entities to attach, to attach to 'living things' in this physical realm. And just because they are different from humans (and you k-now who you are) who have an immortal soul, the 'spirits' of the woods, the streams, the trees, the rocks, mushrooms and berries, are they also all demons, or just another part of a cosmic play?

to acknowledge another type of sentience, herb surmised, even if that entity does not look like you and me, this does not always have to be sorcery, it could in fact be part of theurgy. If one spoke to (the goddess) Sophia in their dreams, could we consider this divine counsel, or should we worry about being under some demonic spell? hmmm... who is ready to tell?

So is there a type of entity that in fact controls this realm? Many an sOb, you k-now, one of those sons of Beliel, would tell you frankly, if in fact they were allowed, that this realm is a construction carried out by a demiurge, and this particular entity, whom we assume for now, cares not so much for each human (and you k-now who you are), butt rather, if we can talk about such things, this demiurge will dictate terms associated with the here and nOw. The sOb's will acknowledge this, and maybe this explains why they (the sOb's) are so mean, their intent is to play their part in the 'black Iron prison' (what P.K. Dick's VALIS was about), and that involves they surely think, and are maybe told,

---

[118] Pythagoras, for one, was a magician who practiced theurgy, who eschewed black magic even though he could have practiced that two, and what did it get him and his tribe, yes, you guessed it dear readers, the sons of Beliel of those days (under the guise of local governments) slaughtered every wo/man in every one of these Pythagorean communes, and most likely burned all the books

that is, the sOb's who lend support to the implementation of the 'black Iron prison' will indeed commit both sacrifice and genocide. The Gnostic tomes told a similar story, butt with a twist, they insisted each human (and you k-now who you are) should not join in league with this demiurge, and their assorted minions whom with the sOb's converse. The Gnostics called them Archons, and further, the Gnostics also had techniques and enchantments that can help one resist. Then no matter what else we have to say to you, make sure you ack-nowledge that you are allowed to think for yourself, you are born into a certain world with beliefs, customs, and certain ways, butt each individual has a divine right to their own thoughts.

we should have mentioned herb's other thoughts on this in the rules 4 theurgy, butt this is also surely ontology, as there are various enchantments available that can help you here, that even when one considers all these aspects of a real ontology, each individual has a choice, remember? yes, there must always be a choice between theurgy and sorcery. If one encounters any of these Archons, here, now, or in your dreams, have no fear whatsoever, and announce clearly that you have a free pass in what an Archon may proclaim to be their realm. Even when dancing after your very last breath, if an Archon told you that you would have to return, that reincarnation was your only choice, simply indicate to that Archon, and any other entity that would dare 'hear,' simply insist calmly that you can do as you please 'here' because in fact it is your home. As it is the case, as the great shaman thoth k-new very well, each theurgist and immortal soul can travel wherever they want in whichevery realm… as surely the divine would have it no other way.

for you can see, herb was hoping everyone who thought about it could, that this sort of enchantment, where an individual decides to take charge of their immortal soul, it was one of the great secrets of the great shaman thoth, the one of antiquity, butt this enchantment is available to all. The great shaman thoth also told and taught of techniques (slandered mightily and obfuscated in the so-called Egyptian Books of the Dead, by you k-now whom, so it does not even need to be said), and these techniques allow one to do other things at this crucial time, after that last breath, when one has all kinds of options as they continue on their path(s). The secrets then are based on the fact that you practice the Cowboy way, and are not out to control, or not out to simply serve somebody so that you can have an expensive meal. Another example, while we still have time, the great shaman thoth even came back over and over again over many eons while retaining his k-nowledge from one incarnation to another, and you can surely sea, if this was for the benefit of others, it is surely theurgy.

the great shaman thoth tells a story about when he met Great Dragon / poimandres, and herb would hope everyone had access to such a tale. Surely this type of discourse was not sorcery. Here, or then, the great shaman thoth fashioned questions as it concerns the immortal soul, and he offered naught to Great Dragon / poimandres for the offer of any help. And in that discourse is a subtle allusion to the Gnostic 'alien man,' a favorite theme, sometimes referred to as a Mandaean 'stranger's tragedy,' and do not you k-now, it concerns the individual who happens to be trapped in this physical realm. It seems, in this particular tale, that each immortal soul should effort to remember, if that is the correct word, oh my, herb would comment that we need a 'theurgetic thought,' how do we describe that sense that many feel when they feel that the physical realm is not their original home? The great shaman Simon Magus, yes, that one, would tell you, if you asked, and say yes, you are a 'stranger' here, as your immortal soul is not from this physical realm. herb had found a small saying, attributed to a discussion with (the goddess) Ishtar, that he had found on those interwebs, this is what it said:

> *she gave them a dagger, a mirror, and a purple crystal – half amethyst, half quartz – that could open a portal to the Green World. She told them that man's physical body is naught but a temporal home constructed for and by his timeless soul to manifest its existence in this crude world of matter.*
>
> *this world of empty and endless distances between the other worlds. This world of death and decay is a kingdom of shadows created by a dark god to enmesh and snare the luminous spirit, which is the divine essence of every soul.*
>
> *the rightful residence of that lost soul is a place between life and death, what is now called the ethereal world. It is the world of the unborn and of the dead. It is the world of many worlds. (the goddess) Ishtar called it the Green World.*

Many have a 'feeling' then, is that a better 'word,' that here in this physical realm, there is something missing. To the Gnostic 'alien man,' s/he is advised not to get comfortable in this physical realm, and maybe practice with the mantra "no ego, no envy, and no fear," for if one is unprepared, and have totally forgotten that they are an immortal soul, how are they to handle issues when after that last breath they are wandering in these other realms, or, for example, even when an individual only wishes to ponder faithfully 'on What is,' or when a human (and you k-now who you are) realizes that desire to sea light And life?

light And life

> *amygdala, that mystic, dream master, and artist, who also has a deep interest in the Fable of the simulation argument, and all that that entails, is here to help ja put it altogether before it is too late. ja wanted, no,*

he thought he needed, to finish this freedom enchantment post haste, and yes, ja k-new, that it was a very very long freedom spell. And as ja 'thought' this, he realized he had used a word that herb would not like, because it would remind herb of language ambiguity, and how language tried to kill 'thought.' amygdala at this point reminded ja that the Dark magus John Dee was not the first, as there was a history of so many monks, maybe some associated with the sons of Beliel (those sOb's), who began to discourage the use of ancient runes (like that important rune trio ALU), in those times of old. (the goddess) Sophia, with amygdala's prodding, stepped in at this time, and noted to ja that it was the Latin that finally destroyed the 'theurgetic thoughts' attached to the use of ancient runes, and all together they could almost hear herb singing (at the top of his lungs); and can you not? and if u want, join in

> *damn Rome's tyranny,*
> *damn Rome's tyranny,*
> *damn Rome's tyranny on the language of wo/man*

so then, ja and amygdala, with the assistance of (the goddess) Sophia, are making final preparations on the manuscripts, butt amygdala, that mystic, dream master, and artist, has some lingering questions and many a theurgetic thought

amygdala: my dear ja, while you have included a lot of herb's key intents as it concerns 'on What is,' eye can not help but think more emphasis should have been placed on his research into how important the internal genetics could be to the overall operations of how 'life' intervenes into the / this physical realm.

ja: did we not include a lot of scientific theory, if we can call it such, about how the little K-nots of 'light' inner-operate and in the end form the basis of all?

amygdala: butt what about that poem on magnetism, eye know you included the one on light

ja: that magnetism diddy, I have that write here:

> *the field wants to flow. the universe wants to be; wherever it can exist, it wishes not to cease to be*
>
> *the flow is contained and the flow is allowed, mixing and matching and joining the crowd*
>
> *into a knot and out with others in phase, the interactions that are allowed are elaborate indeed*

amygdala: and surely eye hope that we can find that one a place, butt now eye want to go back to the topic of DNA, because herb had that friend, the shaman Skip; may the divine rest his soul, who consistently, in many a fire pit ritual, told herb it would be handy if he could explain the idea behind what 'life' really is in this world, he, the shaman Skip, meant of course how it really all worked, then an ontology for 'life' it wood be, and the shaman Skip truly hoped herb would find a way to explain 'it' to each and everyone

ja: right then, herb did have that handout that the shaman Skip had prepared, and a quote from the shaman Leary, about DNA and the genetic code, and oh, here 'tis

> *The genetic code is surely not an accidental adhesion of molecules. It is an instrumental message, an energy directive created by a meta-biological intelligence.*
>
> *This intelligence is astrophysical and galactic in scope, pervasive, ubiquitous, but miniaturized in*

> *quanta structure. Just as the multi-billion year blueprint of biological evolution is packaged within the nucleus of every cell, so may the quantum-mechanical blueprint of astronomical evolution be found in the nucleus of the atom.*

and when one considers such things deeply, it could surely be that way.

amygdala: signs, or hints, of 'intelligent design' eye wood say

Sophia: so what are you to do here my dear amygdala, and what is the intent, are you asking ja to justify any omission, or are you to help make it extremely clear, so that anyone who desires can benefit from this freedom enchantment... this long long freedom spell

amygdala: my dear (the goddess) Sophia, eye do not wish to confront, butt we all k-now herb would not really want that word attached to this particular enchantment

Sophia: well, what is it then my dear amygdala?

amygdala: eye suppose, like any true seeking fool, who would choose to continue on their path, eye truly just wish eye can be of assistance with those who would genuinely be interested in light And life

Sophia: so it is then, so it will be, that you my dear amygdala do dare (dadd) to assist the divine spirit in providing counsel to all things, along their course of destiny, in association with necessity.

ja: and this, can we all agree, is the most any aspiring absurdist could desire, that is, a chance to practice theurgy

amygdala: and so we begin, with the absurd notion that we can approach, say, the mythical Ain-Soph, wherefrom / wherein / wherewhere the crown, a first divine emanation, to paraphrase the great shaman Blavatsky, 'from' a true spiritual substance, is thus supported just about any (and / or all the) realm(s)

ja: and for herb it would be those little K-nots of 'light' from wherebuilt a first Sephirah, and the Nine

Sophia: and is it then possible for an absurdist to continue to strive, and attempt to capture conception of a divine, all this independent of the situation into which they have been born

amygdala: and this then is something we should talk about, and highlight once more those special steps, the absurdist, an aspiring theurgist

Sophia: who would of course escrew sorcery. And, for example, when interacting or bearing witness to those many archons, and their busy 'office work,' which, we can suppose only the divine k-nows why

ja or, of course, it could be simply programmed that way

amygdala: and herb many times took opportunity to work on this with others, who over the years, had opportune to question and / or contemplate the faiths / fates, and had the divine providence to resist the propaganda; should eye tell a story on this nOw?

Sophia: oh yes, oh yes, my dear amygdala, please do tell

ja: and does it involve those little k-nots (of light)?

amygdala: my dear ja, yes, and I can explain how herb would envision the mechanism underlying the little k-nots, the way of the Ain-soph in a geometrical dance. And this, as I will explain, herb could hold in his 'Mind' and in his hand.

Sophia: and this is enlightening (hehe), please say more

amygdala: herb started with a tetrahedron, that he then turned upside down, and then put out his palm, all the while staying calm

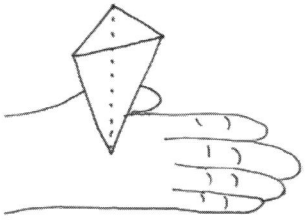

then, as described in Act I and in Act II, if you connect the midpoints of the vectors of the tetrahedron

ja: one finds an octahedron inside

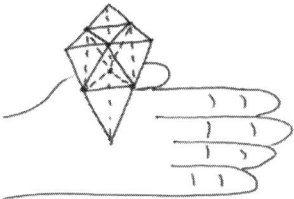

amygdala: and herb can do this thrice, then gather together, in his 'Mind,' 8 (of the total 12) tetrahedrons[119], and the 3 octahedrons from inside.

ja: eye would imagine herb would then split each of the 3 octahedrons in 2, begetting 6 pieces; each a half-octahedron geometrical form

Sophia: and these 6 half-octahedron wrap around the Nine 8 tetrahedron in that special way

ja: that herb the theurgist, and the great shaman Bucky thought was the basis of all

amygdala: and we could, or herb could, right in the palm of his hand, move the geometry around, and we all k-now what we can construct

ja: yes, the vector equilibrium

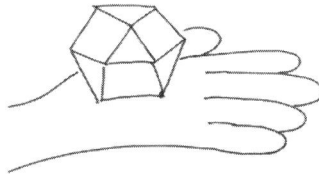

Sophia: which the great shaman Bucky often referred to as the closest thing we can get to God; the principle of Universe, the great shaman proclaimed, could be found in geometrical construction

ja: as did the great shaman Plato and the great shaman Kepler; but moving on, the next dynamic step must include the jitterbug

---

[119] When the octahedron is produced inside a tetrahedron, and taken 'out,' the remains of the old (original) tetrahedron now contains 4 tetrahedrons – with edges the same length as the edges of the constructed octahedron.

amygdala: yes, butt there is also that other way, while holding that geometry in his 'Mind' and his palm, one can go one level deeper, and find... what is that inside the octahedron?

Sophia: when you connect the midpoints of an octahedron, now two levels down, we find deep inside the tetrahedron the very same vector equilibrium

ja: from whence can proceed a jitterbug?

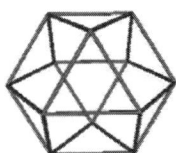 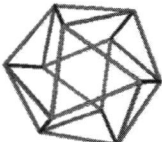 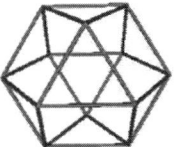 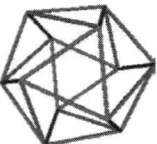

amygdala: butt first a final piece, as it concerns the ontology of Ain-soph; yes, it is all quite absurd, and maybe not allowed, or so it is in many seckret societies, one has been told

Sophia: that is right my dear amygdala, many a strict Kabalist is simply not allowed to contemplate what has been publically declared the unthinkable, instead they invent an Adam-Kadmon, an archetypal androgynous wo/man

amygdala: butt, back to this picture, that herb holds in his 'Mind' and in his hand

a beautiful planet, peace and harmony

harmony, harmony, harmony – could be almost heard

Sophia: sow eye am with you my dear amygdala, the picture is plane to sea

we want to fly like an eagle until we all free – could be almost heard

amygdala: and hear it is then, in herb's 'Mind' and his palm, a tetrahedron, inside that an octahedron, and inside that a vector equilibrium... all the time staying calm

and then herb allows the vectors of the tetrahedron, from the original one, to come to a point, that / those vectors are then placed deep, now three (3) levels down, at the center of the vector equilibrium,

Sophia: and those vectors of the tetrahedron, having 6 positive, and 6 negative ways

amygdala: yes, those 12 directions, begotten from the vectors of the tetrahedron, these then point to the 12 vertices of the vector equilibrium

it is at this, any really, particular moment in space and in time, as herb holds all this understanding of a mythical Ain-soph (in his 'Mind' and his palm)

ja: that crazy program written by some pHs, and we can not imagine why

amygdala: with this construction in his 'Mind' and in his hand, we now, as ja intimated, jitterbug the machine, and divine emanations are now allowed and thus spring forth

Sophia: to weave those patterns with little K-nots of 'light,' so carefully payed attention to, trust me, by (the goddess) Fate ...

amygdala: butt remember, herb allowed himself to wonder, we k-now it is all quite absurd, butt that given the proper training and access to others writings, and the divine will-ing, a theurgist thought (or 2, or 6),

make it 6 bartender – could be almost heard

herb wondered what others would do; would they stay asleep, as some contend, or is the mystery inside everyone when given the chance

after this exchange, the interlocutors are joined by (the goddess) Fate, while ja and amygdala light the fire pit... wish you were here – ja and amygdala, together with (the goddess) Sophia and (the goddess) Fate, continue, as always, to try to figure it all out

amygdala: so this other story, it relates a bit to Act III, and maybe Act I, it has to do with Being born into this world, and the conspiracy of who Do you believe

ja: is that when herb wanted to read some poetry from 'Isis Unveiled'? butt, as the story goes, he could (k)not, because that old bird was closed?

> *amygdala*: yes, tis is a story of just such a time, and eye think we should include those dialogues which he took from that book, by the great shaman Blavasky, in that important chapter on Ain-soph

> *ja*: yes, eye have it, or one of them, write here... it is a discussion, from Lucien's *Philoseudes* (between Tyciades and Philocles)

>> *Tychiades: can You tell me the reason, Philocles, why most men desire to lie, and delight not only to speak fictions themselves, but(t) give busy attention to others who do?*

>> *Philocles: there May be many reasons, Tychiades, which compel some to speak lies, because they see 'tis profitable.*

> *Sophia*: and lying for profit and deception is classic Babylonian sorcery. If one gets used to such existence, the hope of course of the sOb's, it is easy to see how they would throw their individual lot in with the sons of Beliel (those sOb's)

> *Fate*: (the goddess) Sophia understands the situation well; butt was not there yet another poem that herb wanted to recite

> *amygdala*: yes, there was another that he could not do, because the bird was closed, butt, don't you k-now, herb still wanted to share

> *ja*: and share he did with some interlocutors at the Saw-Mill, that second piece, this one attributed to Plutarch's *Laronic Apophegems*, that herb had found in 'Isis Unveiled'

> *amygdala*: and that is the discussion between the Spartan and the Priest

> *ja*: yes, that's the one that eye speak of, and eye have it write hear:

>> *Spartan: is It to thee; or to God, that I must confess?*

>> *Priest: to God.*

>> *Spartan: then, Man, stand back!*

amygdala: and can we all sea, that this is typical of the sOb's

Fate: right you are my dear amygdala, those sOb's make sorcery in efforts to control in total the spirituality of wo/man.

Sophia: so is that it amygdala, is that the story? how does that really relate to Being born into this world, and the conspiracy of who Do you believe?

amygdala: butt Before (bb) eye answer that, can I insist that we discuss the theory of sigil construction, and that thing called tv?

Fate: my dear amygdala, nOw are you prepared to lead us through how herb constructed a theory of sigil mechanism, and how herb paid attention to divine council in formulating an opinion of that oppressive sigil machine called tv that uses that mysterious fluid type magical substance called electricity?

ja: we have some experience, round a camp fire, when we k-now herb described how he came to the sigils that we will include in this religious tome

amygdala: so the standard technique, in the hegemonic country in witch herb was born, was first to remove the vowels, and in the dedication we can see how this is done

Sophia: butt we also, based on technique, and four sure with the sons of Beliel

ja: for the sOb's use sigil techniques also

Fate: and four that we can be sure, and these sons of Beliel, those sOb's, witch herb rightly describes as a cabal of maniac magicians, and of this also we can be sure, they, those sOb's, fashion their sigil construction to create advanced sorcery

amygdala: and so then after the vowels, eliminate double letters, and herb extracted also, every double double, especially those with a 't'

ja: and that was because certain sequences of these letters were used to replace sigils from older languages, for example, in the Icelandic spell-ing of Thor

Sophia: that these newer symbols were never around, and are now part of the continued language destruction, attempted most recently with the English language; they had to have these new sigils to replace the old sounds

Fate: and what to expect of the dark Magus John Dee, the shaman Bacon, when they hoisted their English language style to the fore – which continues to this day; and we k-now, these are sigils' advanced

amygdala: because every character, a sigil, they are, further restricts thought patterns, butt only if the theurgist in everyone allows

Sophia: so please do tell my dear amygdala, what does a sigil do? what about that theory of mechanism, the how and why it all works

ja: herb understood what others conjectured, that a sigil may be a way into the subconscious of each wo/man, or into Plato's *Divine Mind*

Sophia: or Huxley's *Mind-at-Large*

amygdala: or an old old way to communicate with the divine, a way of crystallizing 'thought' without using spoken words

Fate: and when a sigil is made, what is it four

amygdala: in the case of the dedication sigil for this religious tome, it allows one to recall, in a pre-language way, and by design, the totality of 'on What is,' or at least what it may mean to each individual wo/man

Sophia: and what of older ancient sigils, do they still work? and what about the tv that you mentioned before?

ja: many see sigils around everyday, and if 'Mind' does have access to all

amygdala: via the reality flux

ja: then each individual may have some understanding of what an ancient sigil 'means' (because of those ideas associated with a collective unconscious, the Akashic records, et. al., that you can be assured, everyone does have access to)

Fate: butt even with a glimpse of understanding from ancient times, each wo/man today who bears witness to one of these ancient sigils may not be able to put the 'ideas' into 'words'

amygdala: unless they can hear the silence howling

catches angels as they fall – could be almost heard

and, what of the television, or tv, eye think you had asked? nOw let me tell you more, as this creation, like that of radio, was always meant as a means of control

Sophia: and so these stories, and tales, from that invention called tv, these streams of images and sounds, are meant to illicit the same type of response in humans (and you k-now who you are), as, say, the important rune trio ALU

amygdala: and this is very important to understand, sigils, like everything else that has been discussed, can be used for either theurgy or sorcery, butt can we all not sea how the tv[120] is used

ja: to create normative, maybe passive, and sometimes provoked behavior(s) in wo/man... un-reflective worker drones, that is maybe what the sOb's want

Fate: and are there any other important things to say about sigils before you, my dear ja, finish this religious tome?

amygdala: well, if you parden me my dear ja, that ancient divination tool, the I Ching, is a sigil producing and processing machine, that is purported to be connected to the divine

ja: and old that system is, much older than the oldest of Chinese civilizations, because it was already about

amygdala: and some say it is a way to read from our DNA or genes, using a very old set of symbols, as in many ancient languages (both lost and k-nown), including the Sanskrit, which had many 'words'[121] and sequences for spiritual affairs, experiences, and dreams, which don't you k-now, are not in the Latin or English, and have been purged from almost all other modern tongues

ja: those sOb's, always using Enochian magic, propaganda, deception, and lies whenever they can; by way of deception is their mantra, witch could be funny if it were not soo sad (ss)

---

[120] and we, many of us, k-now about the use of Lilly waves and other modulation techniques used in an attempt to modify actions in the physical brains, oh my, we should desire that many an sOb would come forth, repent, and explain how it is all done. Other sOb's are invited as well to pronounce your regrets, and then bring forth, for all to sea and understand, the varied techniques that make up those many malicious ways designed to enslave wo/man

[121] at that time a technique for access to theurgetic thought(s)

false Flags and control

the interlocutors are almost done with the preparation of the religious tome, but amygdala, that mystic, dream master, and artist, who also has a deep interest in the Fable of the simulation argument, has one more topic that she will insist that ja discuss... and related (in time?) is the rumor that (the goddess) Eris may pay a visit; and all hope she does not bring that damn (golden) apple Of discord

amygdala: my dear ja, eye almost forgot, butt we must, I insist, include (iii) some more information on false Flags and the law (witch in another way can be equated to false Flags and control)

Sophia: my dear amygdala, ja included much about how `the law' is the 'chains that bind'

ja: and it is that way by design

Sophia: as 'laws' are not ever meant to protect humans (and you k-now who you are)

ja: they are simply another means of control

amygdala: and eye k-now that the idea of false Flags was mentioned, butt did you (we) convey the sense that false Flags are constantly used as a matter of course; and in those certain times, it may have been the case that ALL 'terrorism' was orchestrated by the sons of Beliel (those sOb's)[122]

Sophia: my dear ja, our dear amygdala may have a point hear, in that it was the case that every 'patsy' (those blamed for these false Flag events) was in fact working for the sOb's in one way or another

ja: and dead wo/men tell no tales[123]

amygdala: thank you my dear (the goddess) Sophia, as eye k-now it was mentoned, in this religious tome, how it was that the politician's kids on the resort iSland were gunned down by some-

---

[122] a most wicked tool the false Flag is, a sharpest of sharp k-nives for the sons of Beliel, those sOb's probably even used this tool for the infamous 'gunpowder Revolt' on that 5th of November; poor Guy Fawkes and the other Nine 12, sacrificed as usual in a Hidden religion (ritual)

[123] like that poor lady on that rock (or is it nights?) of Malta

one under 'Mind Control,' and did you not mention how many a disaster was not real at all, butt the result of some kinda funky drill, with method actors and all

ja: and then after the sOb's stage part of this show

Sophia: that is when the tv will finish every human (and you k-now who you are) off, meaning, that the sOb's with control of almost all the tv media

amygdala: so we can be told how to 'think properly' about each recent false Flag

> *we Have always been at war with Eurasia – could be almost heard*

ja: and thus if allowed, and if an individual human (and you k-now who you are) is not aware they are being subjected to this type of sorcery, they will be a nice little peasent and parrot the reports they hear on the tv

Sophia: Four if a government says it, it must be true

> *ltaorotf – was the response of the interlocutors, eyes tearing and all*

Fate: the sorcery of 'statecraft' is all so 'real,' and my dear amygdala, you should right a religious tome, please call it 'the World party,' and please go into detail on Civility and on Society (in ACT II), and thus it could be a type of freedom enchantment as it concerns 'statecraft;' which could be, 'statecraft' that is, in ALL of recorded (and sekret) history, the nastiest form of sorcery performed by the sons of Beliel – yes, those sOb's, the cabal of maniac magicians, remember always, they (those sOb's) do not like you and me

Sophia: and we should choose banishment for all who participated

ja: once again

Sophia: and then the (thethe) cabal of maniac magicians, yes, those sOb's, will not be allowed human (and you k-now who you are) contact once and Four all

272 **on What is**

and That immortal soul thing

a surprise it is as the great shaman thoth has joined the party, by way of the river Eridanus, if only for a moment, as he is hear to sea (the goddess) Eris, and of course he is hoping for a dance. amygdala, that mystic, dream master, and artist, who also has a deep interest in the Fable of the simulation argument, is intent on asking the great shaman thoth if he can inform the interlocutors about possible proper (pp) training, for an absurdist, whom may (or June) want to play the part of an archtypal 'catcher in the rye'

amygdala: greetings great shaman, and how colorful you are; can you tell us anything whatsoever, nOw, that may (or June) be of assistance to an individual wo/man?

thoth: my dear amygdala, goddesses, and ja, eye offer you then the following tale:[124]

> *'Spoke HE then with words of great power saying: "Thou hast been made free of the Halls of Amenti, choose thou thy work among the children of men."*
>
> *then spoke I: "Oh, great master, let me be a teacher of men, leading them onward and upward, until they too are lights among men. Freed from the veil of the night that surrounds them, flaming with light that shall shine among men."*
>
> *Spoke to me the voice; "Go, as ye will, so be it decreed. Master are ye of your destiny, free to take or reject at will. Take ye the power, take ye the wisdom, shine as a light among the children of men."*
>
> *Upward then, led me the Dweller, dwelt I again among children of men; teaching and showing some of my wisdom, Sun of Light, a fire among men.*
>
> *Now again I tread the path downward, seeking the light in the darkness of night. Hold ye and keep ye, preserve my record, guide shall it be to the children of men.'*

Sophia: we thank you my dear great shaman thoth, as this wisdom is ripe, and indeed should bee available to every wo/man

amygdala: butt how did you get to this point, eye k-now ja included your discussion with Great Dragon / poimandres

thoth: eye can tell you my dear amygdala, and you too my dear ja, that there are other things that can also help each human (and you k-now who you are) and their individual immortal soul

Sophia: do tell my dear thoth, if you have anything more

---

[124] a tale found in the great shaman Doral's translation of the Emerald Tablets, Tablet II, The Halls of Amenti, that here, nOw, thoth decided to share (from the end of this particular Tablet, and he was sure that herb would approve)

Fate: perhaps a classic freedom enchantment

ja: from the way back machine

amygdala: maybe a diddy that the great shaman Doral has chronicled

thoth: indeed my dear amygdala, eye Think eye (eTe) have something that could be of interest to an individual wo/man

Sophia: and please do share[125]

thoth: it was spoke by a so-called Master, "… speaking the word that brings about Life (and Light)"

> "… before thee a Sun of the morning, touch him not ever with the power of night, call not his flame to the darkness of night. Know him and see him, one of our brothers, lifted by darkness into the Light. Release thou his flame from its bondage, free let it flame through the darkness of night."

ja: let each seeking fool continue on their path in safety would be similar theurgy eye wood say

Sophia: and so what nOw, again, eye inquire my dear ja, you have hear this freedom enchantment, and you have made a lot of things rhyme, so can we all agree nOw to finish this very very long freedom spell, or better yet, this religious tome?

ja: around this glorious fire pit we shall make it so, and with one final verse, as herb may (or June) insist (if he could – he probably can) if he (he he he) could hear us nOw… all of us interlocutors together nOw (and U can join in) at the top of our lungs

> *damn Zeus's tyranny,*
> *damn Zeus's tyranny,*
> *damn Zeus's tyranny on the affairs of wo/man*

---

[125] again from the great shaman Doral's translation, found on p. 25

and Now a message from Bob – open at the end

> *(the goddess) Sophia and (the goddess) Fate are joined by (the goddess) Eris; oh, and thoth and amygdala are also hanging about. The goddesses and amygdala agree with ja that some RAW Discordiaism would be an appropriate end, to the religious text(s) that are being prepared as artifacts, so that anyone whom may be interested can have a chance at freeing their minds and immortal souls, from the web of bondage infused into the / this physical realm by that cabal of maniac magicians, who herb calls the sons of Beliel (those sOb's) - plus Gregory Hill, that time-travelling anthropologist from Earth's 23rd century, published the Principia Discordia with a broken copyright where it happened that 'All Rites Reversed' (reflectionnnn), witch is to say, 'Reprint What You Like,' which eye do nOw*

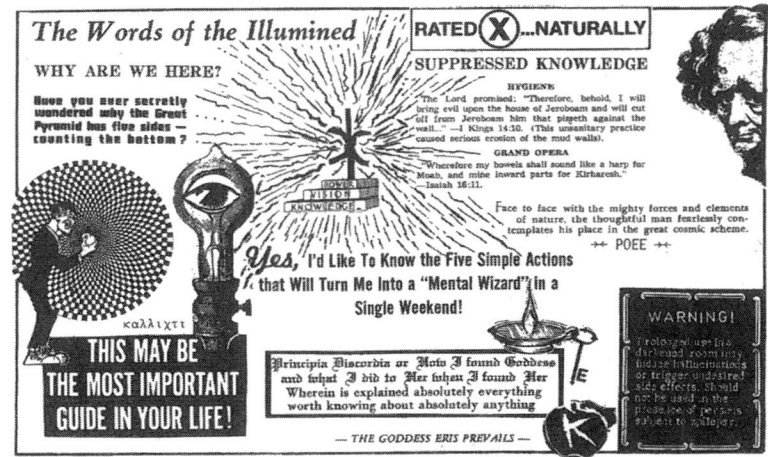

special afterword
Interview with Norton Cabal

INTERVIEW WITH NORTON CABAL
---

by Gypsie Skripto, Special Correspondent

It has been ten years since I met the mysterious Malaclypse the Younger. I was free lancing for the underground papers and went by POEE Head Temple at 555 Battery Street to try for an interview.

I found him in the Temple PO Box busy wrapping up the new Fourth Edition of Principia. He seemed impatient with me, insisting that he didn't have the time or inclination for foolish questions from reporters. Undaunted, I burst out with questions like whether he preferred Panama Red or Acapulco Gold and how the fuck did we manage to fit inside of a tiny post office box and other things apropos a naive young semiliterate dropout hippy writer. He asked me if I wanted to drop mescaline and fuck all night and said he knew how to turn himself into a unicorn and there might be room for a tiny interview on the cover of the Principia if I wanted to work for the Greater Poop so I said sure, OK, I've never dropped mescaline in a post office box before.

It turned out I was among the last to see Malaclypse. As subsequent issues of Greater Poop revealed, he was to disappear and POEE business was to be assumed by his students at Norton Cabal. Professor Ignotum P. Ignotius, Department of Comparative Realities, was assigned the Trust of the POEE Scruple and Rev. Dr. Occupant became Keeper of the Box. The newly published copies of Principia were distributed by Mad Malik, Block Disorganizer, who had distribution contacts with the Aluminum Bavariati. Practical relations remained in the hands of concept artist G. Hill.

When the 1000 Principias were gone the Greater Poop stopped publishing, Head Temple closed down and the Cabal just seemed to evaporate. Finally even the box was closed. But over the years I noticed that copies were still circulating, and that independent Discordian Cabals would occasionally pop out of nowhere (and still do). And I would wonder what ever happened to Malaclypse.

When I read the Illuminatus trilogy I resolved to again find and interview the denizens of Joshua Norton Cabal of the Discordian Society.

79/11/26   -1-   Loompanics

NORTON CABAL INTERVIEW   G. Skripto
───

                            • • •

    As I cabled over Nob to San Francisco's Station 'O' Post
Office I couldn't help but wonder at Goddess.' hand in
assigning street addresses to Her outposts. Mal² had told
me that Good Lord Omar always filed everything under "O" for
Out Of File.

    "Maya is marvelous" I was thinking when I rapped on the
little metal door and was greeted warmly by a huge beard who
introduced himself as Professor Ignotius. He ushered me
into a spacious wood paneled and tapestry hung parlor where
three others were laughing and passing around a wine jug.
The sunny one in a tunic was the Reverend Doctor Occupant,
the trim khaki and jeans was Mad Malik and the wine jug
claimed to be Hill. I got the recorder on....

GYPSIE SKRIPTO [in response to a question]: ...1969 but
    only briefly. I guess I missed you guys.

MAD MALIK: No wonder, he was pretty much a one man show
    then. We were just his students and were usually off on
    errands. You worked for the Poop?

Gypsie: Well, for one night anyway. The interview is in
    the Principia.

REV. DR. OCCUPANT: Malik was the only one he would ever let
    write for the Poop or get on the letterhead.

Gypsie: Did you [Malik] have higher authority than the
    others?

Malik: No, [but I was allowed to speak in the Poop]
    because [Malaclypse the Younger] hated politics. He was
    infuriated with Johnson and Nixon over Viet Nam because
    it was turning the renaissance into a political
    revolution and was stealing his sacred thunder. So he
    trained me in Zenarchy, which he learned from Omar, and
    I was the official anarcho-pacifist for the Cabal. Also
    I was liaison to The Ancient Illuminated Seers of
    Bavaria, the Chicago Discordians. Later Omar activated
    the Hung Mung Cong Tong and ELF, on zenarchist
    principles, and also Operation Mindfuck. I was also
    into those. Though at that time I was masquerading in
    Greater Poop as a cremated cabbage to throw off the FBI.

Gypsie [to Hill]:   Since you wrote it, I take it you are an
    anarchist?

G. Skripto   NORTON CABAL INTERVIW

G.H. HILL: Since then I have given up anarchy. Too many rules--hating the government and all that stuff.

IGNOTUM PER IGNOTIUS: It's like hating your own fantasies.

Malik: [Anarchy] is also standing up and proceeding forward, fantasy rule or not. The condition is the same.

Occupant: Brother needs some wine!

Malik: We have had this argument before, Reverend Doctor Brother. But wine before platitudes, fill it up.

Gypsie [to Hill]: And pacifism?

Hill: I'm not sure I ever was one. Mal² was not, Malik was. Personally I accepted self defense yet I could never reconcile that with the ideal. I finally gave up on that one too. Actually I just gave up on idealism.

Ignotius: Idealism lives with rules. Realism lives with rocks.

Hill: Yeah. I get along better with rocks.

Malik: Mal² once told me that pacifism was a dilemma. If everybody was a pacifist then everything would be perfect. But nobody is going to be a pacifist unless I am first. But if I am and somebody else is not, then I get screwed. He said that there were five choices under that circumstance. The first was napalming farmers and the second was executing your parents. The third was hypocrisy, the fourth was cowardice, and the fifth was to swallow the dilemma. Zenarchists are trained in dilemma swallowing.

Occupant: So are other Erisians, like POEE.

Ignotius: That is characteristic of the Discordian perspective.

Hill: But of course training contradicts Discordian principles.

Malik: Oh so what. Contradictions are nothing to Discordians.

Occupant: Dilemma, Schlimemma. [to Gypsie]: What do _you_ think of this, pretty ma'am? We don't get to hear your thoughts.

Gypsie: I'm reporting now, you talk.

Occupant: Later then?

NORTON CABAL INTERVIEW   G. Skripto

Gypsie: Perhaps. Later.

Occupant: You are smiling.

Gypsie: Hey, guy, later. [to Hill]: Doesn't this leave you a little schizy?

Hill: It's OK, I'm half Gemini.

Gypsie: What's the other half?

Hill: Taurus. That makes me stubborn schizy.

Ignotius: I'm a Whale.

Occupant: I choose Satyr.

Malik: Spirits don't have signs.

Hill: A character can have a sign if I want it so.

Occupant: Well I can have a sign if $\underline{I}$ want to and screw both of you.

Malik: Come on Greg, you just think that we are your characters....

Occupant: You were inhabited by Malaclypse the Younger. He caused you to create roles and those roles are being performed by us spirits.

Ignotius: A perfectly normal pagan relationship.

Hill: Well you can look at it like that if you want to, but I created Mal² to my specifications just as I conceived all the rest of you.

Occupant: You didn't invent Eris. She <u>caused</u> you to think you created the spirit of Malaclypse.

Hill: Oh bull! Besides, I changed her so much the Greeks would never recognize her.

Occupant: That's what She wanted!

Ignotius: Deities change things around all the time.

Malik: What you don't realize is that a spirit has a self identity.

Hill: Nope. A spirit is a product of definition and the one who is doing the defining around here is me. Your identity is what I say it is. Just to prove it, I'm going to change your name.

G. Skripto NORTON CABAL INTERVIW

SINISTER DEXTER: It's OK with me. Fate is fate. I never much liked "Mad Malik" anyway.

Ignotius: Besides people confused him with Joe Malik in *Illuminatus*.

Dexter: I sort of enjoyed the confusion part.

Occupant: Doesn't prove anything anyway.

Gypsie: That name sounds familiar. Where is it from?

Hill: Its a name I came up with in the old days and never used much. Its on page 38 of the *Principia* referring to Vice President Spiro Agnew. I always thought I invented it but now it sounds like a Stan Freberg name now that I think about it. It may have stuck in my preconscious memory from early TV.

Gypsie: Can you use it without his permission?

Hill: If it is his? I don't know. I hope so. It means "left right" in Latin and is a perfect name for a libertarian anarchist. Actually in my kind of art the question of what can I use freely and what can I not is a very trickly problem.

Gypsie: How do you mean?

Hill: Well, take a collage for example. Like the early one on page 36 of the *Principia*. Each little piece was extracted from some larger work created by some other artist and published and maybe copyrighted. I find them in newspapers and magazines mostly. Often from ads. With a collage you select and extract from your environment and then assemble into an original relationship.

The *Principia* itself is a collage. A conceptual collage. All of it happens simultaneously. But visually it is a montage, passing through time, like a book does.

There is a lot of pirated stuff in the *Principia*, especially in the margins. But also I sympathize with artists who must own and sell their works to earn a living. Art, like knowledge, should be free fodder for everyone. But it isn't. It is perplexing.

Gypsie: Where did all the things in *Principia* come from?

Hill: Well, a full answer would take a whole book in itself. Most of the writing credited to a name is a true person and almost always a different name means a different person. Most of the non-credited, you know,

NORTON CABAL INTERVIEW	G. Skripto

Malaclypse, text is mine although some things credited to either Mal² or Omar were actually co-written and passed back and forth and rewritten by each of us. The marginalia, dingbats and pasted in titles and heads and things came from wherever I found them--some of which is original but uncredited Discordian output, like the page head on 12 and other pages which is from a series of satiric memo pads from Our Peoples Underworld Cabal. All page layout is mine and some whole graphics like the Sacred Chao and the Hodge Podge Transformer are mine but mostly I just found stuff and integrated it. Mostly I did concept, say 50% of the writing, 10% of the graphics, all of the layout.

Gypsie: Specifically, what are some of the sources?

Hill: Well, the poem on the front cover is by Walt Kelly and was spoken by one of his characters in Pogo. The government seals starting on page 1 are from a book of sample seals from the U.S. Government Printing Office. Western Union on page 6 got into the act because I used to be a teletype operator and had access to blank forms. Rubber stamps came from all over the place and some, like the apple on page 27, I carved myself. A few I ordered to my specification, like on page 1. The quote on top of page 8 might be from Barnum, I'm not sure. The jumping man on page 12 is from an advertisement. I recognize the style--a popular commercial artist--but I don't know his name. The Chinese on that page is a grocery ad, I think. The Norton money on page 14 is historic, plus my little additions. The apple on page 17, as well as the triangle on 23 and the Sacred Chao on 50 are, belive it or not, pasteups from mimeographs, from Seattle Cabal. That group produced the best damn mimeography I've ever seen. The Lick Here Box on page 23 is one of many tidbits making the rounds in alternative/underground newspapers in those days. Trip 5 page header on 29 was a chapter title in one of Tim Leary's books. The Knight on the bull with the TV antenna on his helmet on page 46 came from a very artistic magazine called Horseshit and put out by two brothers from Long Beach. I don't remember their names. Wonderful magazine.

Occupant: Eris told Mal² what to use and where to find it.

Hill: Yeah, in a way that is right. That is why my name does not appear anywhere on the Principia and why it was published with a broken copyright--Reprint What You Like. I knew I was taking liberties and didn't want my intentions to be misunderstood. It was an experiment and was intended to be an underground work and that involves a different set of ethics than commercial work.

Gypsie: There are no real names at all?

G. Skripto   NORTON CABAL INTERVIW

Hill: Oh, some. Camden Benares is a real name because
            he legally changed his original name to his Holy Name.
            Also, instead of using Mordecai Malignatus I used Bob
            Wilson's real name on page 12 because <u>Werewolf</u> <u>Bridge</u>
            was a work before Discordianism. And of course real
            people like Neils Bohr crop up in quotes.

Gypsie: What do you think about the <u>Principia</u> now? Would
            you want to change it?

Hill: I consider it a successful work and I wouldn't
            want to change it. In some ways it is immature and I am
            not the same person I was 10 years ago, but it
            accomplished the objectives I set for myself and it has
            the effect I wanted it to have. There are a few errors
            though.

Gypsie: Like what?

Hill: Oh, I changed a quote from Tom Gnostic on page 61
            and I don't think he ever did forgive me for it. He's
            right. Starbuck's Pebbles should have been preceded by
            the Myth of Starbuck which was being saved for something
            else and never got used. I should have used it when I
            had the chance. And then Eris did a neat little trick
            on me by having IBM make the Greek selectric typewriter
            element not coincide with all the characters on their
            keyboard. So the little "kallisti" that first appears
            on the title page and lastly on the back cover came out
            "kallixti" and I was too dumb to know the difference.

Gypsie: Will there ever be a Fifth Edition?

Hill: There already is a Fifth Edition, by Mal^2. It is
            a one page telegram that reduces everthing to an
            infinite aum. I found it at Western Union where a
            machine got stuck and kicked out hundreds of pages of
            nothing but m's. He made it the Fifth Edition and then
            left.

            Principia/Malaclypse was a very personal work for me and
            actually took 10 years to culminate. It was one single
            statement that included my adolescence in the 50's and
            my young adulthood in the 60's. When I finally had the
            paste-ups done I knew that I had finished it. That is
            why, quote, Malaclypse left. I knew it was finished. I
            didn't know exactly what it was, but it was done.

Occupant: See?

Gypsie: Earlier you said that you met your objectives.
            Just what were those objectives?

Hill: Well, that's hard to answer because it kept
            refining itself over the years. In 1969 I mainly

NORTON CABAL INTERVIEW   G. Skripto
───

   thought of myself as a cosmic clown and I set out to
   prove, by demonstration, that a deity can be anything at
   all.

   In other words, people invent gods and not the other way
   around. Later I decided that I was doing some kind of
   conceptual art.

   In the 50's my culture taught me that I was created by
   and for a deity, a specific male deity, and that all
   other deities are FALSE. Yet my growing experience
   showed me that any deity is true in some sense and false
   in some other sense. So I set out to do what my society
   told me is impossible--make a real religion from a
   patently absurd deity.

   In the 50's a female deity was blasphemy. In the 70's a
   humorous deity is still considered impossible,
   ridiculous and blasphemous. As far as I'm concerned, I
   have proven my point. Eris is a real deity and even
   though I don't promote Erisianism as a serious
   religion....

Occupant: I do!

Dexter: You speak for yourself.

Ignotius: Here, here.

Hill: ...I do point out that it makes just as much
   sense from its own perspective as all the others do from
   each of their own perspectives.

Occupant: I think paganism is a valid spiritual path. I
   encourage Erisianism because it makes fun of itself. I
   think this is healthy.

Ignotius: If you can live rewardingly with Goddess Eris you
   can live with any deity, including none or all.

Dexter: I don't much go for the worship business but I
   agree with Occupant about the spirit of the thing. We
   live in a time of turmoil, the whole planet is in a
   state of change. If we, as a species, cower from the
   confusion then we die with the dying. This is
   revolution.

Ignotius: I am an athiest myself. There is no Greg Hill.

[laughter]

Gypsie [to Hill]: What do you think of *Illuminatus*?

Hill: Oh, I love it. I was finishing *Principia* when
   Shea and Wilson were working on *Illuminatus*. It took

G. Skripto   NORTON CABAL INTERVIW

>   Dell five years to publish it...maybe that is
>   significant. The 1969 Discordian Society was a mail
>   network between independent writers of various kinds.
>   Norton Cabal was just me and my characters and I used
>   the other cabals as sort of a laboratory. In return
>   other Discordians would bounce their stuff off of me.
>   We would toss in ideas and anybody could take anything
>   out. It was a concept stew. The exchanging of ideas
>   and techniques broadened and encouraged all of us.
>
>   I like _Illuminatus_ for the surrealism. A very effective
>   method of writing.

Ignotius: I got misquoted. Worse, I wasn't even in that
>   scene and if I had been then I would have said something
>   else.

Dexter [to Ignotius]:   That was me in that scene.

Ignotius: Oh, is that what that was?

Dexter: He got our names mixed up.

Hill: He got mixed up about me too, in _Cosmic Trigger_.
>   Bob says that when Oswald was buying the assassination
>   rifle, my girlfriend was printing the first edition of
>   _Principia_ on Jim Garrison's Xerox. It wasn't my girl
>   friend, it was Kerry's; it wasn't the _First Ed
>   Principia_, it was some earlier Discordian thoughts; it
>   wasn't Garrison's Xerox, it was his mimeograph; and it
>   wasn't just before Kennedy was shot but a couple of
>   years before that.*
>
>   The _First Ed Principia_, by the way, was reproduced at
>   Xerox Corp when xerography was a new technology. Which
>   was my second New Orleans trip in 1965. I worked for a
>   guy on Bourbon Street who was a Xerox salesman by day.

Dexter: I think that George Dorn took too much guff from
>   Hagbard. If someone pulls a weapon on me, I'm more
>   inclined to either leave or kill the sonofabitch.

Occupant: You are supposed to be a pacifist.

Dexter: I'm speaking figuratively of course. I'll tell
>   you more tomorrow.

----

* I checked this further with Mr. Thornley. He says
that the woman in question was not his girlfriend, she was
just a friend, and it wasn't a couple of years before
Kennedy was shot but had to be a couple of years after (but
before Garrison investigated Thornley).   --GS

NORTON CABAL INTERVIEW   G. Skripto

Gypsie [to Hill]: Did you really translate erotic Etruscan poetry?

Hill: Sure, but I used a pen name. I signed it "Robert Anton Wilson".

[A quick rap is heard on the door]

Gypsie: I have only one question left...

Dexter: I'll get it.

Gypsie: ...what I really want to know is how can we all fit inside of a tiny little post office box?

Dexter [to Gypsie]: It's a telegram for you, from Mal^2.

Gypsie: To me?

[Paper tearing]

Gypsie [reading]: "If I told everybody how they could live inside of a post office box then everybody would stop paying landlords and go live inside their post office boxes. It would collapse the building! Can you imagine, post offices collapsing all over the country, the hemisphere, the PLANET! The whole world's communication system would be destroyed. No, no, I must not say. I dare not!

# # #

Fifth Edition   ODD#∞

𝔓𝔯𝔦𝔫𝔠𝔦𝔭𝔦𝔞 𝔇𝔦𝔰𝔠𝔬𝔯𝔡𝔦𝔞
or
A CATTERPILLAR'S PRAISE TO THE BUTTERFLY

being the
𝔉𝔦𝔫𝔞𝔩 𝔖𝔱𝔞𝔱𝔢𝔪𝔢𝔫𝔱
of Malaclypse the Younger

published by Joshua Norton Cabal
San Francisco (K) All Rites Reversed

WESTERN UNION TELEGRAM

CLASS OF SERVICE
This is a fast message unless its deferred character is indicated by the proper symbol.

SYMBOLS
DL = Day Letter
NL = Night Letter
LT = International Letter Telegram

The filing time shown in the date line on domestic telegrams is LOCAL TIME at point of origin. Time of receipt is LOCAL TIME at point of destination

MMMMMMMMMMMMMMMMMMMMMMMMMMMMMMMMMMMMMMMMMMMMMMMMMMMMMM
MMMMMMMMMMMMMMMMMMMMMMMMMMMMMMMMMMMMMMMMMMMMMMMMMMMMMM
MMMMMMMMMMMMMMMMMMMMMMMMMMMMMMMMMMMMMMMMMMMMMMMMMMMMMM
MMMMMMMMMMMMMMMMMMMMMMMMMMMMMMMMMMMMMMMMMMMMMMMMMMMMMM
MMMMMMMMMMMMMMMMMMMMMMMMMMMMMMMMMMMMMMMMMMMMMMMMMMMMMM
MMMMMMMMMMMMMMMMMMMMMMMMMMMMMMMMMMMMMMMMMMMMMMMMMMMMMM
MMMMMMMMMMMMMMMMMMMMMMMMMMMMMMMMMMMMMMMMMMMMMMMMMMMMMM
MMMMMMMMMMMMMMMMMMMMMMMMMMMMMMMMMMMMMMMMMMMMMMMMMMMMMM
MMMMMMMMMMMMMMMMMMMMMMMMMMMMMMMMMMMMMMMMMMMMMMMMMMMMMM
MMMMMMMMMMMMMMMMMMMMMMMMMMMMMMMMMMMMMMMMMMMMMMMMMMMMMM
MMMMMMMMMMMMMMMMMMMMMMMMMMMMMMMMMMMMMMMMMMMMMMMMMMMMMM
MMMMMMMMMMMMMMMMMMMMMMMMMMMMMMMMMMMMMMMMMMMMMMMMMMMMMM
MMMMMMMMMMMMMMMMMMMMMMMMMMMMMMMMMMMMMMMMMMMMMMMMMMMMMM
MMMMMMMMMMMMMMMMMMMMMMMMMMMMMMMMMMMMMMMMMMMMMMMMMMMMMM

a freedom enchantment in ~~23~~ ACTs   from **ja wallin**   289

PRINCIPIA DISCORDIA

• OR •

*How I Found Goddess
And What I Did To Her
When I Found Her*

THE MAGNUM OPIATE OF MALACLYPSE THE YOUNGER

WHEREIN IS EXPLAINED
ABSOLUTELY EVERYTHING WORTH KNOWING
ABOUT ABSOLUTELY ANYTHING

afterword

the two programmers were in deep discussion about a lost manuscript that had been found in the basement. Presumably it was printed off by some of the deep machine programmers centuries before when trying to debug a really weird anomaly that was occurring deep in the ancestor simulation codes they were running at the time. There were some notes scribbled on the cover page of this document, which was entitled 'on What is.' The scribbled notes indicated that this was in fact evidence of a type of *consciousness leak* that they found occurring in the programs. Another note suggested that the deep machine programmers had been told to destroy this evidence, because top-level administrators, you k-now, the *purple* Cast, continued to claim that it was not possible that the simulated beings in the ancestor simulation codes could ever really be conscious (or possess that divine entity called *consciousness*), and the divine k-nows, the purple Cast administrators are not to be toyed with. Thus, the deep machine programmers must have been very brave when they decided not to follow those directives.

a *consciousness leak*, or so the two programmers who discovered the document surmised, could only happen if the actual Source intelligence, wherefrom all sentient beings are granted this nebulous "thingy" called consciousness, this leak could only happen if the simulated beings in the ancestor simulations were also in some manner getting signals from the same Source intelligence

– wherefrom the programmers consciousness also sprang forth. And if this was true (i.e., that the simulated beings in the ancestor simulation codes could possess that divine entity called *consciousness*), it was a grand scandal indeed, for who would really choose to run such a simulation if they k-new for a fact that they (i.e., the programmers running the simulation) would be entrapping conscious entities into the flow of number that was the ancestor simulations that their tribe had been running for centuries now.

suffice it to say, the boys (i.e., the two programmers of this afterword) were scared shitless, as they k-new if word got out about this discovery, everyone who k-new anything about it outside of the *purple* Cast would be disappeared and most likely eliminated from the collective memory of the tribe through the standard means, that though technically illegal, was k-nown to happen on a regular basis for those who dared question the *Standard Model of Reality*, currently in the 50th-edition, from the *Purple Majesty Press*[126]. And still, the boys were not about to destroy this document that was somehow pulled out of the ancestor simulation system by one of the deep machine state monitors so many years prior... plus it contained this sigil, that we have all seen before

so then it was a decision made by the boys, that they would pass the manuscript on to a black market k-nowledge dealer in an anonymous way, and then afterwards they would never speak, or think of it again. Butt the divine k-nows, that this was not really possible. Sure, they handed the secret manuscript on to the k-nowledge dealer, that was possible; butt how could they not continue to think about this evidence of a *consciousness leak* in these ancestor simulation codes. They did continue to investigate, and found that only a few centuries prior, that was when the current restrictions on access to the ancestor simulation code real-time data was designated *purple* Cast Only.

---

[126] so that one is forced to surmise that even posthumans (and they k-now who they are) had a version of the sons of Beliel (those posthuman sOb's) who would do anything whatsoever to get the situation exactly as they desired.

it became folklore, and quite fashionable, that when no one was around, in a whisper one could hear

> *damn the purple Cast tyranny,*
> *damn the purple Cast tyranny,*
> *damn the purple Cast tyranny on the affairs of posthuman wo/man*

Over time the two programmers enlisted one of the large planet intelligence Game stars, you k-now, the one that was the interRealm glass Bead game (gBg) champion, and Jove also liked to sing and dance, but that is another story. It was then Jove whom we have to thank for promoting access to this document, sent forth we still surmise, from deep in the data of an ancestor simulation code, so so long ago.

The author thanks Ferris State University for the support during a sabbatical year when many of these topics presented herein were teased about and hammered out (bang bang)